Panentheism

Addressing

Einstein

And

Imaginary Numbers

Daniel J. Shepard

Resolving the Paradox Regarding

- **Space and Time**
- **Imaginary Numbers ('i')**
- **A System Built Upon Time and Space**

- **Nonfiction / Metaphysics / Ontology / Systems Theology**
- **Second Printing (July 4, 2014**
- **ISBN-13: 978-1507741986**
- **ISBN-10: 1507741987**
- **CreateSpace.com**
- **W.E.Hope, Inc. (World Embracing Hope)**
- **1. Philosophy 2 Panentheism 3. Space 4. Time 5. Imaginary Numbers 6. Title**

A gift

From me to you

From one soul to another

Peace

The Gift

Note to the reader:

- The intent of the more than 20+ books is to provide enough material to prove the validity of panentheism not beyond 'all' doubt but to prove the validity of panentheism beyond 'all reasonable' doubt. The point being to elevate the individual's and our species' perception of themselves in order to elevate human behavior on both an individual level and on a species level before we begin to step into the heavens.

- The series of books, Panentheism, emerged from earlier metaphysical editions and have been edited and retitled to more accurately reflect the true nature of their contents.

- I understand there are numerous stylistic, grammatical and spelling errors within all my work. I hope you as a reader can overlook such issues and focus upon the ideas being presented. I do not like to make excuses but most the material is, after all, free (see panentheism.com) to the public and therefore producing no revenue stream.

  Having spent more than a quarter of a million dollars on the web site: panentheism.com, 20+ books, presentations, videos, attempts to place the material in the hands of academics and the public, ... I found my resources insufficient for formal editing. It is perhaps best to consider the products of my work more as a personal log in the rough of what it is I have been entrusted, with the condition that I pass this material on to you.

Daniel J Shepard
Channel
Panentheism.com

Books by Daniel J. Shepard

Panentheism

Vol. 1: Panentheism addressing Humanity's Purpose
Vol. 2: Panentheism addressing Man made in the Image of God
Vol. 3: Panentheism addressing Sci./Rel./Phil./and Prophecy
Vol. 4: Panentheism addressing Volumes 1 – 3 Guide
Vol. 5: Panentheism addressing The Physical and the non-Physical
Vol. 6: Panentheism addressing Humanity Confined to a Universe
Vol. 7: Panentheism addressing Free Will and Determinism
Vol. 8: Panentheism addressing Anthropocentricism
Vol. 9: Panentheism addressing Theodicy
Vol. 10: Panentheism addressing Universal Ethics
Vol. 11: Panentheism addressing The Lack of 1st Cause
Vol. 12: Panentheism addressing Einstein and Imaginary Numbers
Vol. 13: Panentheism addressing The Mathematics of non-Members
Vol. 14: Panentheism addressing Creation from the Void
Vol. 15: Panentheism addressing Monism/Dualism
Vol. 16: Panentheism addressing Nihilism
Vol. 17: Panentheism addressing Language
Vol. 18: Panentheism addressing Philosophy's Responsibility
Vol. 19: Panentheism addressing Ockham's Razor
Vol. 20: Panentheism addressing Panentheism
Vol. 21: Panentheism addressing On being 'the' Summit
Vol. 22: Panentheism addressing Do we need to change? Can we change?
Vol. 23: Panentheism addressing Western Philosophy
Vol. 24: Panentheism addressing Chaos/Complexity
Vol. 25: Panentheism addressing Abbreviated Thoughts
Vol. 26: Panentheism addressing The Whole of Reality
Vol. 27: Panentheism addressing The Soul
Vol. 28: Panentheism addressing God/Brahma
…

More information can be found at my web site

❖

www.panentheism.com

Panentheism, a small seed planted into the social fabric of our species. An idea which only takes one Greek word to express, 'panentheism' and three English words to explain, 'pan' all, 'en' in, 'theism' God. 'All in God' and with that simple phrase our species has the potential to change forever.

djs

Project Overview

1995 - 1996 Final draft of "You and I Together: Have a purpose in reality" completed. This was a process of coalescing forty years of thoughts regarding a Universal Holistic System. From these notes, a model was constructed. The impact was then examined regarding this particular model and the effect it would have upon humanity in terms of the most cherished concepts embraced by the individual as well as those embraced by our speciess.

1996 - 1997 Final draft of "In the Image of God" completed. This step involved testing the practicality of a Universal Holistic System. The work examines the ability of the System to resolve twenty futuristic socially-divisive issues and ten current socially-divisive issues.

1997 - 1998 Final draft of "Stepping up to the Creator" completed. Once the system had been developed, the impact examined, and the practicality tested, the Universal System needed to be formalized, expanded, and validated against what it is we believe - religion, what it is we observe - science, what is we reason - philosophy, and what it is we've been told about change - prophecy. The work takes on a three-dimensional matrix format. The matrix format was used to help the reader move in and out of the 900 various topics and levels of difficulty.

1998 - 1999 Final draft of the Cross Reference Guide and Index" completed. Because of the expansiveness of the project, the need arose to find a means of cross-referencing the intricacies of the project. This was accomplished through the development of a cross-reference sectioned into five categories: Questions Addressed, Flowcharts, Thematic Index, Index, and Glossary.

1999 First draft of CD completed: The project was converted into Adobe Acrobat format. This was done to make the project user-friendly. The CD assists the exploration of the project through the power of the search engine called Adobe Acrobat. The CD will be updated as the project progresses.

1999 First draft presently unfolding on site of "On 'being' being 'Being'" This is a technical work intended for deep thinkers. Its intent is, through constructive criticism, to examine the error of humanity's perceptual journey generated by philosophers over the last twenty-five hundred years. The Universal Holistic System of Panentheism acts as the foundation of the constructive criticism.

1999 First draft of CD completed: Multimedia presentation of the www.wehope.com project as well as other misc. lectures. This series of lectures/presentations is made in person. Even philosophers must strive to apply practical applications to their work. The W.E. Hope Foundation is a nonprofit organization established by this philosopher in an attempt to apply the fundamental principles he espouses.

1999 CD - Part I. Audio readings of articles. The CD's are custom made. Please link to www.wehope.com for additional information.

1999 CD - Part II. Audio readings of articles. The CD's are custom made. Please link to www.wehope.com for additional information.

2000 Multimedia Presentation - A Universal Philosophy. This is a 981-slide presentation, in Adobe Acrobat format, that explores the means by which we could attain a universal philosophy. This presentation will be available for online viewing later this year.

2000 In the articles section of the Library page, a number of articles are available for viewing. These are works-in-progress and are intended to be incorporated into a new trilogy to be completed later this year.

2000 A new page "Reflections" has been added to the site. These are an account of my thoughts and reflections on a variety of philosophical issues and questions.

2000 A new page "Aphorisms" has been added to the site.

2000 A new page "Definitions" has been added to the site.

2000 - 2003 The final volume of the third volume of a new trilogy was placed online. The complete trilogy - The War & Peace of a New Metaphysical Perception - introduces a new perceptual model of reality. The work is intent upon establishing the understanding of a new metaphysical system, which combines the Aristotelian metaphysical system of Cartesianism and the Hegelian metaphysical system of non-Cartesianism into one system. The three volumes of the new trilogy are as follows:

2001: Volume I - On 'being'

2001: Volume II - On 'being' being

2001: Volume III - On 'being' being 'Being'

2003 – 2005 Existence: In and of Itself - Introductory Work to Trilogy II: The War and Peace of a New Metaphysical Perception.

2004 Convert and place on line: The War and Peace of a New Metaphysical Perception to an Ontological Version.

2004 – 2005 Convert complete site from HTML to CSS / DHTML to stabilize site for the long term and to facilitate removing and reinstalling site if it becomes corrupted through use or hacking.

2005 New Site Appearance, Complete All Sections of the site except 'Latest Additions', Add additional sections to the site, and Complete Final Appearance of Site.

2005 - 2008 Move the project to the more advanced interactive www tool of blogging: Adding reason to faith URL:

http://panentheism.blogharbor.com/

2009 Development of a new series: Understanding …

2010 Understanding Reality: The four absolute truths secularists are intent upon eradicating are: 1. A Creator of the physical universe exists. 2.The true essence of the individual is made in the image of this Creator and is thus, by definition, divine in nature 3. The individual and our species exist temporarily in the physical for a reason. We have a purpose. 4.The void, ex-nihilo, creation from non-existence did occur. These four fundamental, absolute truths will be addressed in great detail within this book and will, beyond all reasonable doubt, be shown to exist as absolute truths. The theists need more than faith to establish their positions in this day and age and this work gives them what they need to rationalize their positions.

2011 Converting the work into a format compatible to createspace.com and kindle.com. Placement of work onto createspace.com and kindle.com.

2012 Understanding Reality
2013 Understanding the Soul
2014 Understanding God/Brahma

Panentheism

… is the only understanding of reality rationally capable of fully addressing the interrelationship between the individual, time, space, imaginary numbers and what it is our physical universe is expanding into if indeed our universe is expanding as scientists suggests.

Daniel J. Shepard

... in the nineteenth and twentieth centuries, science became too technical and mathematical for the philosopher, or anyone else except a few specialists. Philosophers reduced the scope of their inquiries so much that Wittgenstein, the most famous philosopher of this century, said, "The sole remaining task for philosophers is the analysis of language." What a comedown, from the great tradition of philosophy from Aristotle to Kant!

Steven Hawking, A Brief History of Time
A Boston Book, 1988, p. 174

So it is the twentieth century represents the dark age of philosophy for philosophy condemned itself to the same status philosophy gave to God as philosophy embraced Nietsches perception: 'God is dead'. And as suredly as philosophy embraced the death of God so to philosophy embraced the death of metaphysics.

The error: The paradox of 'time and space'

The perception: Einstein moves our perceptual understanding regarding the Kant/Hegel system being filled with 'timelessness and spacelessness' back into the system being filled with time and space. As such, 'time and space', with the help of Einstein, once again have a location within which they can be found. However, the understanding regarding the role of 'time and space' and the role of 'timelessness and spacelessness', as well as the understanding regarding the interrelationship between 'time and space' and 'timelessness and spacelessness' not only remain in a state of confusion but even more disconcerting, the existence of such an interrelationship is not recognized as a significant aspect of the 'larger' system.

It is this state of confusion which will be specifically addressed within this volume.

Understanding Evolving[1]

[1] For the year 2000 CE, Humanity's entry into the 3rd millennium see page 261

Panentheism Series
Volume 12

Panentheism Addressing Einstein and Imaginary Numbers

Resolving the Problem Of Space and Time Imaginary Numbers ('i') And A System Built Upon Time and Space Via Panentheism

Daniel J. Shepard

Channel

Volume 12

Panentheism
Addressing
$E = mc^2$

God
Einstein
Panentheism
And
Imaginary Numbers
('i')

The Paradox Of:

- ***The Square Root of '-n'***
- **The Abstract Equals the Physical**
- **E/m = c(2) where c = d/t (distance/time)**

Daniel J. Shepard
Channel

Table of Contents **Page(s)**

Part I: The paradox of 'i' 19

1. **Introduction** 19
2. **Dimensions** 25
3. **Goodbye concrete, Hello abstract** 29
4. **Real Numbers** 33
5. **The Tunnel of Abstraction** 37
6. **Imaginary Numbers** 55
7. **The Constancy of time verses the Variability of time** 67

Part II: Resolving the issue with a new metaphysical perception 73

Part IIa: The Newtonian 'i' - Velocity Equals Distance Divided by Time 73

8. **Introduction** 73
9. **Expanding knowing revisited** 75
10. **The constant (k) variable** 85
    a. **The 'constant' factor of variability** 85
    b. **The 'constant' variable of physicality** 89

Part I. Hegel introduces the first mirror: Inverse physicality 89

Part II. Einstein introduces the second mirror: The 'i' inversion 101

11. **[d = t]** 103
12. **[1]** 107
13. **[0]** 117
14. **Introduction to ∞ / 1 and 1 / ∞** 123

15. [∞ / 1] **125**
16. [1 = 0 / 0] **129**
17. [1 = ∞ / ∞] **147**
18. [0 / ∞ versus ∞ / 0] **149**
19. [0 = 0] **151**
20. Time and distance both divided by 1 **157**
21. Knowledge: The universal building block **161**
22. The tunnel of perception **181**

Part IIb: The Einsteinian 'i' – The Constant Variable Equals the Square Root of the Distance Divided by the Square Root of the Time **187**

23. Introduction **187**
24. The square root of Einstein's equations: 'i' **189**
25. Einstein's mirror revisited **197**
26. Illusion **209**
27. The 'real' and the 'real' illusion **221**
28.The real **227**
a. Coherency of time **235**
b. Variability of time **241**
29. The 'real illusion' **243**
a. Incoherency of time **253**
b. Constancy of time **261**
30. The 'Taser' **263**
31. What does it mean **269**
32. Preview Volume 13: The Error of Russell **281**

Terms/Concepts

Constancy of consistency
Constancy of sequentiality
Constant of physicality
Coherency of time
Constancy of time
Constant 'k' variable
D = t
Doppler effects of time
Einstein's mirror
Experiential permutations
Hegel's mirror
Illusion
Imaginary numbers
Incoherency of individuality
Incoherency of time
Knowledge
Metaphysical mirror
'real' illusion
Real numbers
Taser
Tunnel of abstraction
Tunnel of perception
Universal building block
Variability of time

Due to problems with the superscript of Microsoft Word Mac various forms of exponential notation are used in this volume. Example: 'c squared' is sometimes written as 'c' with the notations c(2), c2 or the superscript of '2'.

Volume 12

Panentheism

Addressing

Einstein and Imaginary Numbers

- **Physical Time and Space**
- **Imaginary Numbers**
- **A System Built Upon Time and Space**

Part I: The paradox of 'i'

1. Introduction

Newton, Einstein, and 'i' are the key to understanding how it is we get from the location of 'here' to the location of 'there' and the key to understanding how it is we metaphysically go from understanding the direct proportional interrelationship of time and distance to understanding the interrelationship of inverse time being directly proportional to the inverse of distance and then proceed to metaphysically understanding the concept of squaring such interrelationships.

In essence, this volume examines the very relationship of time and distance whether it is in a form of direct proportionality, a form of inverse proportionality, a form of time multiplied by time, or a form of space multiplied by space.

This is a process of stepping onto a surface of quicksand whose depth is indeterminable. The only tangible aspect of this volume is an intuitive sense that the depth of this 'quicksand' will go well beyond Einstein and his concepts of relativity as it applies to metaphysical thought.

This volume, Volume 12: The Error of Einstein, is the most precarious departure from the past volumes found within the work The War and Peace of a New Metaphysical Perception. This volume departs from the 'known' dilemmas/paradoxes of present day metaphysics into the realm of yet to be defined metaphysical paradoxes.

To avoid such a journey, however, is to turn away from the true nature of metaphysics, which is to explore regions yet to be theoretically examined by science itself.

To avoid addressing potentially hypothetical challenges, which a new metaphysical system may 'encounter', to avoid addressing potentially hypothetical dilemmas, which a new metaphysical system may 'suggest', is to show no confidence in the new system itself.

To avoid the inevitable is in essence to shut down the very concept of what a new metaphysical system is required by its very nature to address.

To shun examining the full implications of a new metaphysical system including its impact upon the theoretical is to shun the obligations of the most basic principles of metaphysics itself: 'To thine own self be true.'

And why is the principle ‘To thine own self be true.’ so basic to metaphysics? Principles are so fundamentally basic to metaphysics because it is metaphysics, which deals with the most basic of principles, principles rooted in the purity of truth itself.

So how are we to delve into such an immense project as attempting to understand the concept regarding:

1. Metaphysically understanding the direct proportional interrelationship of time and distance.
2. Metaphysically understanding the interrelationship of inverse time being directly proportional to inverse of distance.
3. Metaphysically understanding the concept regarding the square of the interrelationships expressed in #1 and #2.

To understand the complexity of direct, inverse, and square relationships of time and distance, we will focus upon mathematics and mathematics’ fundamental explanation regarding the relationship between time and distance. If I were a mathematician, the following concepts could be seriously considered for their mathematical soundness.

Since I am not a mathematician, rather than the mathematical soundness of the arguments being the points to consider, one might better focus upon the metaphysical implications of what bits and pieces may emerge from the following examination of mathematics and what clues mathematics might conceivably provide metaphysics regarding an understanding of what lies outside the physical.

Having established a defense for any irrationality which may emerge from the remainder of this volume, let’s explore where reason, fused with mathematics, might take us in regards to metaphysics as we attempt to resolve the puzzling state existing between Zeno’s ‘i’, Newton’s ‘i’, and Einstein’s ‘i’.

The new metaphysical perception which the individual acting within God creates regarding Zeno, Newton, Einstein, relativity, and the modern physics of quantum mechanics is an unusual one to say the least. Modern physics is immersed in the realm of the physical universe. This is as it should be.

What should not be the case however is the perplexing abstractual state of existence within which modern mathematics (the language of physics) and physics find themselves existing. Mathematics and modern physics find themselves immersed within the realm of physicality with no sense of understanding the abstractual significance of the very physical reality they are examining.

Mathematics and physics are in a state of abstractual confusion.

This state of abstractual confusion was not 'created' by mathematics and physics but rather was created by the inability of metaphysics to break out of its state of uncertainty regarding the most fundamental of first truths: 'I am.' 'The universe is.' '1st cause is.'

This state of uncertainty regarding whether first truth is 'I am.', 'The universe is.', or '1st cause is.', once logically hurdled will allow metaphysics to once again lay down a model which can act as a challenge, act as a guide towards which the energies of mathematics and physics may be directed.

Until a theoretical goal is established by metaphysical ingenuity, mathematics and physics will have no beacon towards which they can advance. Without such a beacon, mathematics and physics will have no choice but to visualize each new advance as a step into the blackness of the unknowable which they find surrounding their reality of the physical. Each step will no doubt expand their horizons, expand the very limits of their presently existing physical universe but each expansion will find itself forever being followed by the question: Into 'what' did our expanding universe just expand?

It is Hegel who points the way regarding an examination of the new metaphysical system, an open non-Cartesian Kantian abstractual system powered by a closed Cartesian Aristotelian physical system, the individual acting within God, panentheism introduced by this work:

The War and Peace of a New Metaphysical Perception. It is this new metaphysical system, the individual acting within God which allows us to understand, in the metaphysical sense, the interrelationship between Newtonian physics and Einsteinian physics.

If the new metaphysical system of the individual acting within God aids us in understanding the connection between metaphysical Newtonian physics and metaphysical Einsteinian physics, what then becomes of the 'i'. Is 'i' a grammatical error? 'i' is not a grammatical error. The 'i' is in fact, 'i' not I. 'i' is the mathimatical notation for 'imagniary numbers'.

It is through the process of applying metaphysics to the concept of 'i' (imaginary numbers) that we begin to understand theoretical metaphysics today as opposed to practical metaphysics and metaphysical engineering.

It is through the process of following the trail the concept 'i' marks as it travels through the physics of Newton and then moves through the physics of Einstein that we gain an understanding as to the metaphysical concepts Einstein's introduction of relativity has to offer us as a species of rational, reasoning entities of individuality.

So where do we begin? We begin by examining the most obvious aspect of our reality. We begin by examining what it is we find ourselves immersed within. We begin by examining the realm we call space.

2. Dimensions

The three most familiar aspects of space are the three dimensions: length, depth, and height.

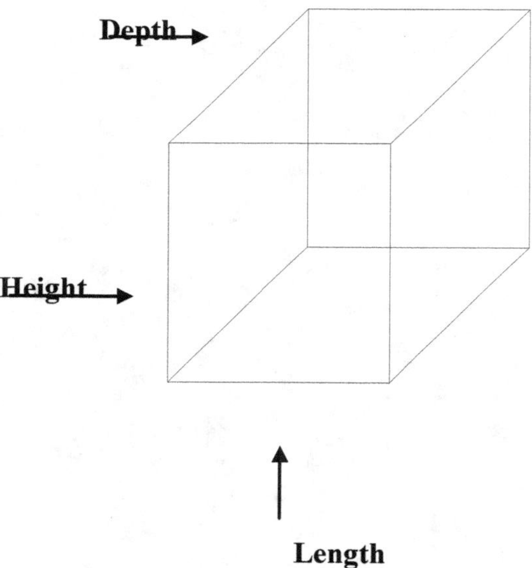

In a sense, dimensions are nonexistent unless 'something', for example: 'matter' and 'energy', are found within the dimensions themselves. Dimensions immerse themselves within matter and energy or one could say matter and energy immerse themselves within dimensions.

There are such 'things' as zero dimensional objects: 'a' geometric point, emotions of love and hate, concepts of ethical and unethical, justice and injustice, knowledge …

There are such 'things' as one-dimensional objects: lines, rays, line segments, open line segments ...

There are such 'things' as two-dimensional objects: circles, squares, triangles, semi-circles, rectangles, rhombi, arcs, parabolas, hyperbolas, ellipses...

There are such 'things' as three-dimensional objects: cubes, spheres, cones, square pyramids, Klein bottles, Mobia strips, dodecahedral objects...

As we introduce more dimensions, we introduce more complexity. Interestingly enough as we introduce dimensions themselves as opposed to the lack of dimensions, we leave the concept of abstraction behind and begin entering the realm of the concrete.

Now this process is fuzzy in the beginning but it begins to come into focus as we move further and further into physical reality through the process of adding additional dimensions.

The point is: Once we have left the concept of dimensions, zero dimensions; we begin leaving the realm of pure abstraction and entering the realm of the physical. This is not to say abstraction no longer exists. Once abstraction exists, how can abstractions ever be erased?

Erasing a picture of a flower does not erase the concept of the flower nor does it erase a flower. Erasing a picture of a flower simple erases 'the' picture of the flower. The concrete item, the flower, and the abstract item, the concept of a flower, remain intact, unaffected by your action of erasing a picture of a flower.

In fact, once the flower has been created, its concept, the abstract understanding of the flower can never be erased for it exists. The question becomes: Which came first the abstract or the physical existence of the flower? Simplified, the question becomes: Which came first, the abstract or the physical?

Metaphysically we continue to come back to the question: Which came first the chicken or the egg.

Darwinian biologists would say the egg came first. But Creationist biologists would say the chicken came first. Again and again the question becomes: Which came first the chicken or the egg? Metaphysically we are no further along than we ever were.

Now I cannot speak for biologists or cosmologists, nor am I suggesting that once the question of which came first, the chicken or the egg, is resolved we will have resolved the question of which came first the abstract or the concrete.

What I am suggesting is that we may be able to better understand the interrelationship between the concrete and abstract if we spend a little time with Newton, Einstein, and i.

So now what? Now we need to get back on task and move to the next dimension, the fourth dimension.

Most of us are familiar with four dimensions: length, width, height and time.

It is within the four dimensions, length, width, height, and time that we find the comfort of our home, our environment, our planet, our universe. It is here we find the sense of belonging as we immerse ourselves in the company of our spouses, children, relatives, friends, coworkers, and fellow humans.

It is here we sleep, eat, reproduce, meditate, contemplate, and vegetate.

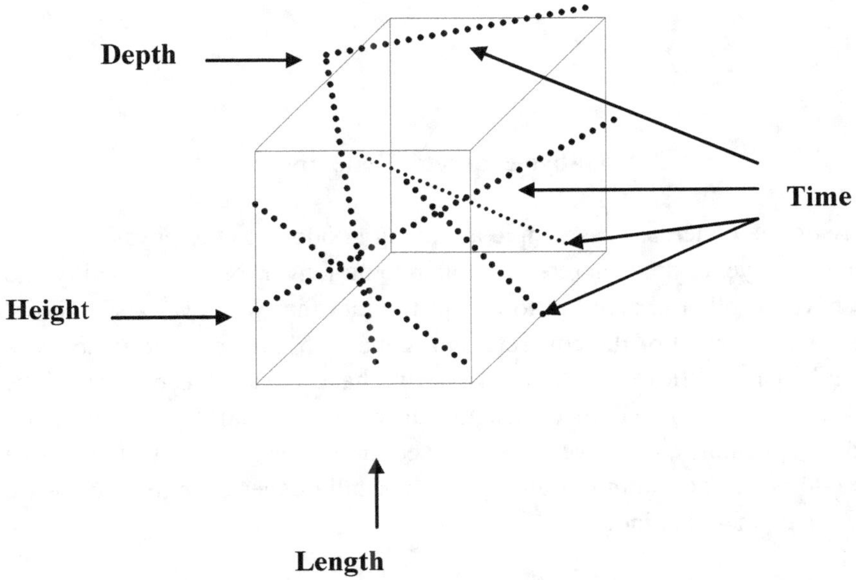

Time cuts through our universe, immerses our universe within itself, and immerses itself within our universe. Time has been shown by Einstein to be a function of what it is we find within our universe, namely matter and energy.

Some would speculate this is not the case. They would say matter and energy are a function of time. This debate is not the issue of this volume, nor is the chicken and egg paradox the issue of this volume.

The issue of this volume, Einstein and i, lies rather in the understanding the relationship of the chicken and the egg, understanding the relationship of time and matter/energy, understanding the relationship of the abstract and the concrete.

So once again, where do we go from here? We need to enter the realm of the abstract. It will be a while before we come back to our home, the realm of the concrete, so if you are a homebody you may wish to skip this volume for we will not be coming home for a long while.

3. Goodbye concrete, Hello abstract

Leaving the concrete is as simple as saying goodbye to the physical. If we say, Goodbye, to the concrete, the concrete no longer exists as a reality and we have no other choice but to accept the fact that we now exist 'within' the abstract. What of the concrete, is it gone? No, the concrete is no more 'gone' than the flower is 'gone' once we have erased the picture of the flower. The concrete still exists, the universe, our galaxy, our sun, our planet, our homes, our loved ones, our communities still exist, its just that they do not exist within our ability to physically experience them once we have 'left' them behind.

Metaphysically the action of leaving the physical behind means we not only leave the physical but we leave the ability to take actions which 'create' new experiences, leave behind the ability to add 'newness' to the whole.

Being within the abstract does not mean we cannot experience what we have never before experienced for we can experience what it is others have experienced, we cam experience what it is we have generated in terms of having created ourselves as entities of unique knowing identified by our unique experiencing generated by our actions of free will applied within the realm of the illusion we call the universe/physical reality.

The complete details regarding the concept of growth of the abstract generated by the active action of free will operating freely within the realm of the physical is addressed in Volumes one through seven.

The objective of this volume, however, lies elsewhere. The objective of this volume is to examine the metaphysical significance regarding the interrelationship between Newtonian physics and Einsteinian physics.

It is because we have committed ourselves to examining concepts of abstract mathematics, because we have committed ourselves to examining what lies beyond the physical/concrete that we have labeled the physical as simply an illusion.

The statement: 'Our universe is an illusion.' needs to be rephrased. It is much too uncomfortable a statement to make so pointedly. It is best for our species that we not make the statement so pointedly since making the statement creates the impression that our universe, our experiences, and we ourselves are simply illusions.

Creating the perception that the concrete is an illusion is not only threatening to our ability to go back into the concrete but creating the illusion that the concrete is an illusion suggests that once we have turned our backs upon the physical, the physical never existed and that goes against everything we believe, reason, observe, or thought we believed, reasoned, and observed.

Once we have said goodbye to the physical, what are we to say then about our new perception of the concrete, which is now beyond our reach other than through abstractual perceptions of belief, observation, and reason.

We must replace the concept of what was, our 'having' experienced the concrete, with the understanding that the concrete 'did' exist, 'does' exist and thus the concrete is not an illusion, but rather the concrete was a 'real' illusion, is a 'real' illusion.

Redefining the physical to be a 'real' illusion as opposed to being simply an illusion helps us understand that once we leave the physical behind and step into the purity of abstraction, we will be able to return to the region we called our home,.

Understanding the concept that the physical remains a location into which we can return may allow us to feel comfortable enough to stay within the purity of our new environment and examine this realm known as the abstract.

Having established a life line back to the concrete, lets now begin examining our new environment, the abstract.

From the point of view generated by the purity of abstraction, we can see various degrees of dimensions. We can see the lack of dimension, zero dimensions.

From the point of view of the purity of the abstract, we can see that our universe is not wrapped 'within' four dimensions but in fact the four dimensions are more than four.

Regardless of how many dimensions we are able to observe, we can see the multiple dimensions, wrap themselves 'around' that 'thing' called the universe, immerse themselves 'within' the universe, find the universe immersed 'within' them, immersed within dimensions, exist as just that, dimensions.

As we look around most of you would drift towards the abstract concepts which personally interest you.

Fortunately or unfortunately, you are with me, a theoretical metaphysician. As such you have no choice but to drift along with me as long as you continue to read this article.

So what then is it I see? I see Einstein's abstract concept: E=mc(2)

I see the c(2). I see c as being squared. I see c: the velocity of light. I see velocity as being an abstraction of distance and the abstraction of time. Furthermore, I see c(2) as being the velocity of light squared, as being the quotient of distance squared and time squared.

I see the abstract from the point of view of the abstract while at the same time I understanding my having experienced of the concrete while having been 'within' the concrete.

I begin to sense a non-contradictory perception regarding the two, regarding the connection of the two, regarding the need of each for the other, of harmony, of brotherhood, cooperation, respect generated by the tool of separation through inclusion as applies to the two, the concrete and the abstract.

However, I am jumping ahead of myself. I forgot for a moment that you are with me and I must go at a pace comfortable for you.

So to begin again: What is it I see? I see a paradox. I see complexity ripe on the vine and waiting to be plucked. I see a situation awaiting the single application of Husserl's reductionism and Ockham's razor simultaneously.

The purity of abstraction provides infinity to the infinite power of possibilities. To make such a vast choice of possibilities manageable, it would be best to examine the infinity to the infinity power of options by examining one concept within this realm of potential perception. The single option of our focus will be the concept of numbers as it applies to Newtonian and Einsteinian physics.

4. Real Numbers

Let's begin our examination of abstraction by examining the concept of Real numbers. The examination of Real numbers is best begun through the initiation of a discussion of Counting numbers.

{Counting Numbers} = {1, 2, 3, …} The set of Counting Numbers is the set of numbers one, two, three, etc. into infinity.

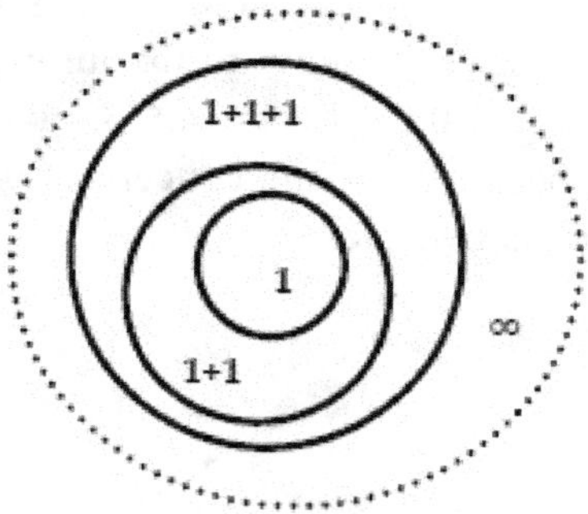

The Counting numbers begin with what is called a 'unit' number, a number of which all the rest are composed. For example the number two is simply one more than the number one. The number three is simply one more two ones

This may not be of interest to most people, but to a theoretical metaphysician it is a truly exciting concept for it, in its simplicity, implies individuality, the individual is no 'little', useless, concept to be discarded for the 'greater' good, the 'greater' idea. In fact the concept of individuality is the bases of all counting numbers, including infinity itself.

Counting numbers lead to the concept of Whole numbers:

{Whole Numbers} = {0, 1, 2, 3, ….} The set of Whole Numbers is the set of numbers zero, one, two, three, etc. into infinity.

A new set evolves, or so it appears.

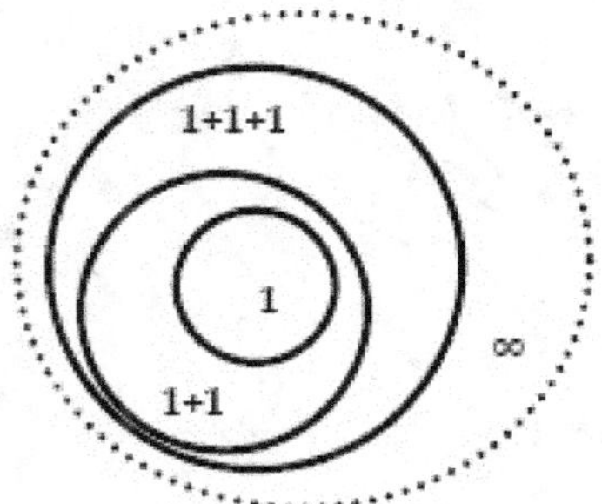

Now why would the phrase be added: 'Or so it appears.'

For one thing, nothing was added to the set of counting numbers. Let me say that again, nothing, literally nothing, was added to the set of Counting Numbers

However, this appears to be the same set as the set of Counting Numbers. That's true. So somehow, we have to distinguish the difference between the two sets. Somehow, we have to 'represent' nothingness being added to the set so that we can understand that 'nothingness' was not a part of the first set, the set of Counting Numbers but is part of the second set, the set of Whole Numbers.

So to distinguish that nothing is the difference between the two sets lets use the symbol: ∅. This will keep our drawings relatively simple.

Now we have:

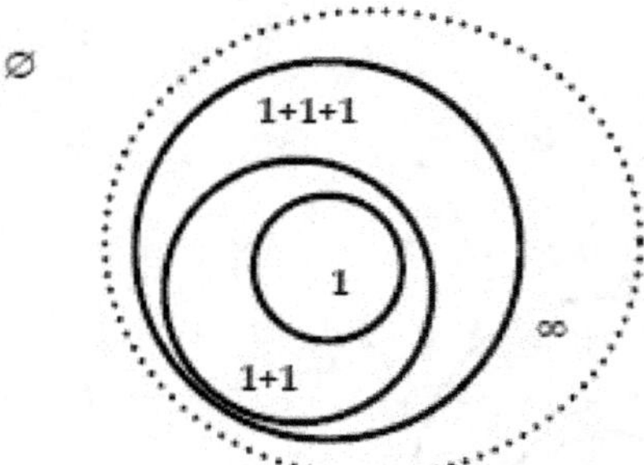

Actually you could have:

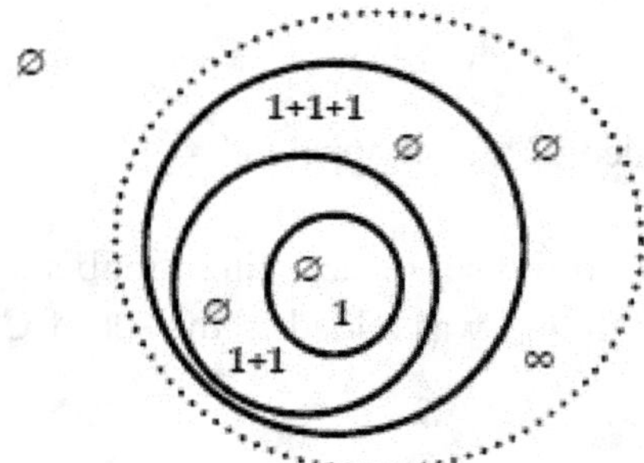

Since ‘nothing has been added.

On the number line however, zero, nothing, shows up once and only once.

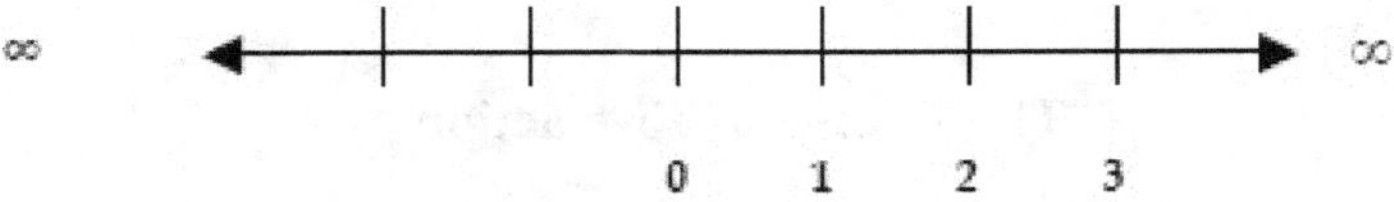

What are the marks on the left of zero, on the left of nothing? We will come back to that very soon. In the mean time lets reconfigure the two dimensional diagram to better represent the one dimensional number line. To reconfigure the two dimensional diagram, we need to develop a three dimensional tunnel within which we in essence could walk and observe the numbers which we pass as we walk through this tunnel.

To avoid getting too complicated too soon lets go back to the set of counting numbers and construct the tunnel and then turn around and walk in the opposite direction through the tunnel.

---

5. The Tunnel of abstraction

Counting Numbers = {1, 2, 3, …}

Tunnel of Counting Numbers:

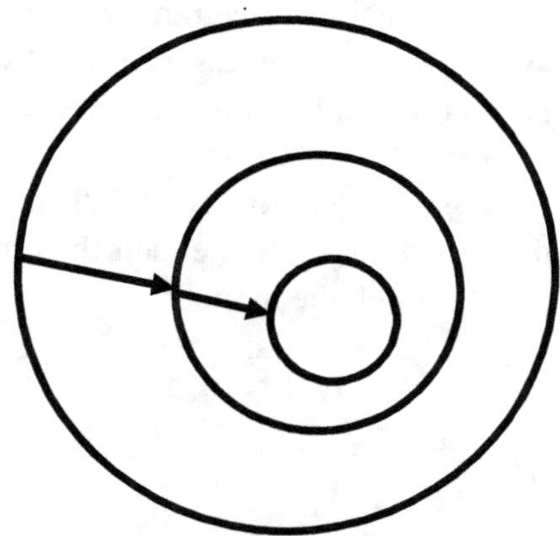

Hmmmm, no the tunnel is going the wrong direction. We must revise this perception for we are individuals, we are the 'one' concept. We, each of us, is 'an' individual. Reversing the tunnel is not a difficult task.

By reversing the concept of the tunnel of numbers we obtain:

Counting Numbers = {1, 2, 3, …}

Tunnel of Counting Numbers:

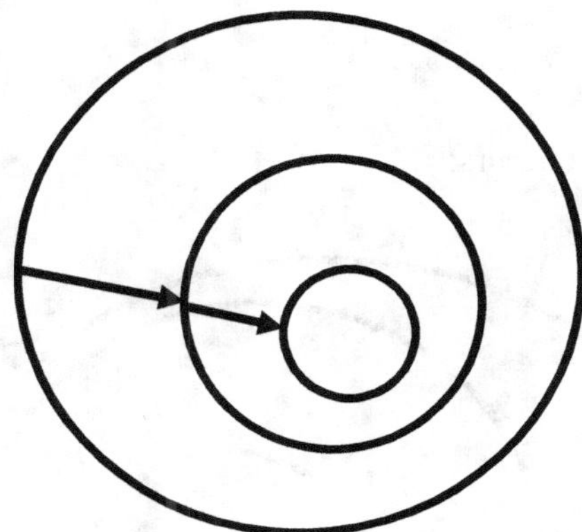

Now we can proceed to the set of Whole Numbers. As we proceed to the tunnel of Whole Numbers, something very interesting develops.

Whole Numbers = {0, 1, 2, 3, …}

Tunnel of Whole Numbers:

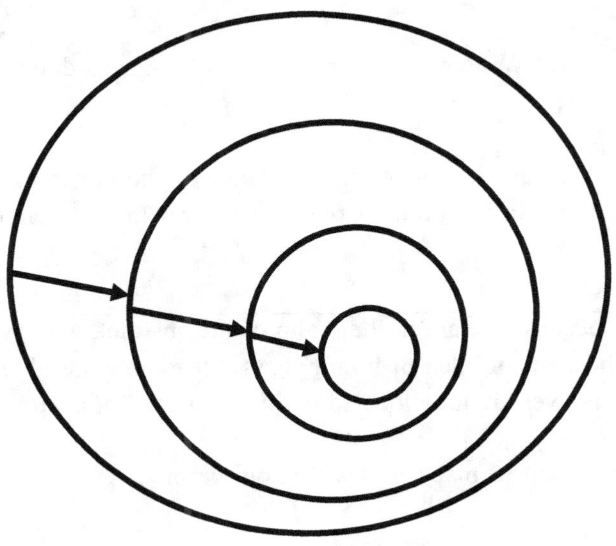

Now 'one' becomes the entity of individuality, as such we must make a further revision to our tunnel of numbers. As we do so, we obtain:

Whole Numbers = {0, 1, 2, 3, …}

Tunnel of Whole Numbers:

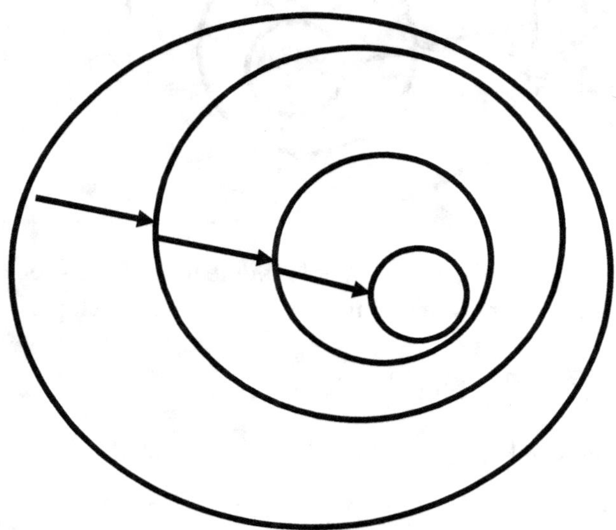

Where is zero? Where is nothingness? Nothingness does not exist in the realm of the physical.

We, however, are no longer in the realm of the concrete. We are in the realm of pure abstraction. In the abstract realm, zero, nothingness does exist.

As such, because we are in the realm of the abstract, we have the ability to understand that the idea of nothingness exists. We are able to understand for we ourselves are now located within the purity of abstraction. A

s such, we can now place zero within our graphic.

In actuality, we have already placed zero, nothingness within our drawing, we just haven't labeled nothingness. To correct this oversight, lets redraw the graphic and include the label 'nothingness'.

Whole Numbers = {0, 1, 2, 3 …}

Tunnel of Whole Numbers:

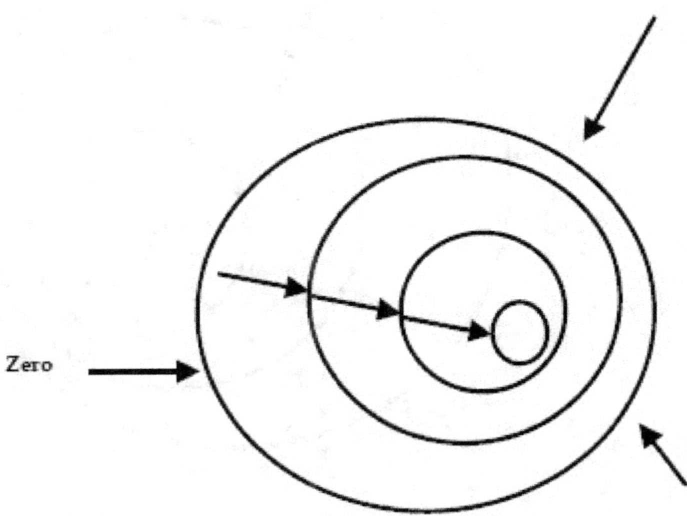

Within this drawing zero appears to take up no 'space' for zero is simply a circle represented by the drawing of a circle and everyone knows lines have only one dimension.

Lines do not have depth or width. This perception is correct. Even in abstraction zero, nothing, is nothing and this drawing implies zero is nothing, only a boundary where one begins, where individuality begins, and moves forward with the concept of multiple individuality.

Expanding upon the concept of numbers and adding the concept of parts, pieces of the whole or what we call positive rational numbers such as 1/2 , 1 ½, 5 ¾,…, we obtain:

Positive Rational Numbers = {#'s > 0 which can be expressed as a/b where a and b are Whole numbers and where $b \neq 0$}

Tunnel of Positive Rational Numbers:

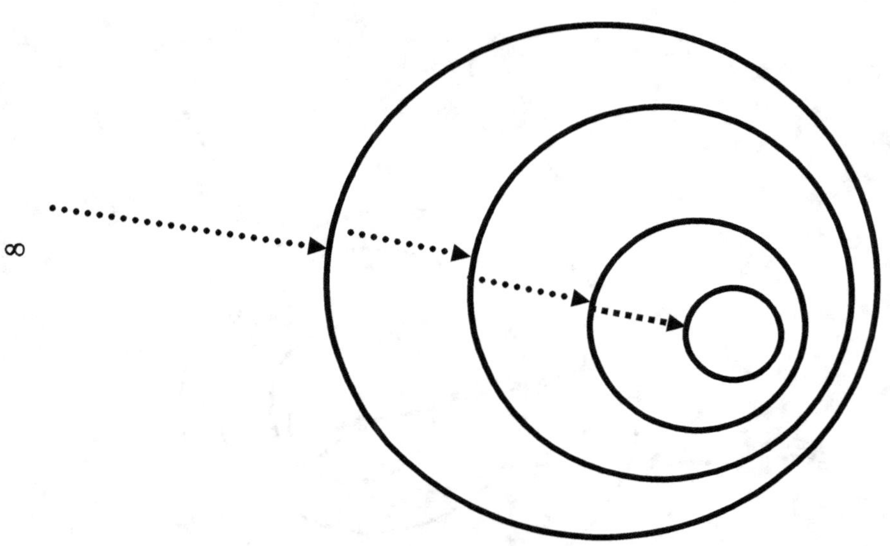

At this point we are going to ignore the concepts of infinite 'largeness' and infinite 'smallness'. Instead we are simply going to consider the concept of size of infiniteness as being simply forms of infinity whose concept of infiniteness alone is what it is we wish to consider. By doing so we eliminate the complexity of $+1/\infty$ as compared to $+\infty/1$.

This graphic leads to the termination regarding ever-expanding concepts of numbers unless we do something regarding nothing. Is doing something about nothing a paradox? Doing something about nothing is only a paradox if we continue our past actions of thinking that nothing has no functionality.

To eliminate the paradox of nothingness having no functionality while remaining what it is, nothing, we must revisit the concept of nothingness and continually remind ourselves that, within the realm of abstraction, all abstractions including nothingness have a function.

The continual need to revisit the functionality of nothingness, leads one towards gaining an intuitive sense that nothingness is an important concept of abstraction.

Reintroduction of nothingness into the graphic gives us:

Positive Rational Numbers = {# > 0 which can be expressed as a/b where a and b are Whole numbers and where b ≠ 0}

Tunnel of Positive Rational numbers now becomes:

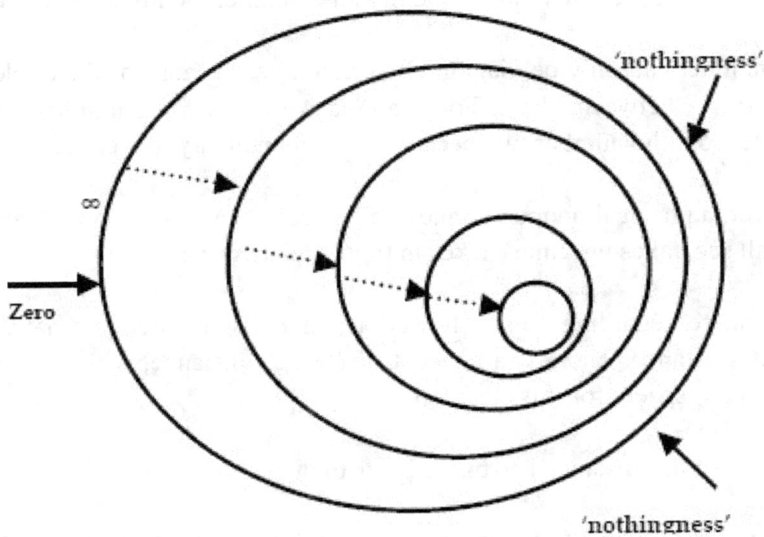

Now what?

Now we move into the abstract realm of the negative. The question becomes:

Is this process of diagramming leading us anywhere? The diagrams are leading us to individuality, the individual.

The diagrams are leading us to an understanding regarding the concept of nothingness itself. Lastly, the graphics may lead us to the very concept of the concrete, being – action, process/reality.

To advance our understanding regarding our present perception of the 'real' illusion of physical reality we next need to deal with the concept of negativity. We need to examine the negative in relationship to the positive.

This leads us to the concept of Rational numbers as opposed to positive Rational numbers. First however there is one more correction, which needs to take place.

As the diagram presently exists, each unique number blends with another.

Each number can only be diagramed as a number because of the circle we have drawn 'between' them. For example we can see the number one is separate from the number two because it has a boundary, the circle.

However, a physical boundary takes up 'space', takes up 'distance', and as we shall see, takes up 'time', takes in time, and incorporates time.

The volume regarding Zeno showed us there is another alternative to concrete distance. There is a concept of abstract distance, which in effect takes up no 'space' for it is abstraction.

Zeno showed us distance has two aspects to it.

Distance has the aspect of physical ness and the aspect of abstraction. Both are real. One is reality when one is immersed within it. From this viewpoint the other becomes a real illusion, but does not 'go away', does not become unimportant, does not become 'just' an illusion.

One can erase the picture of a flower but the flower remains and the concept of the flower remains as well.

Both are real and both remain. Which, the flower itself or the concept of the flower itself, is 'real' is not the question for they are both 'real'. One is 'real' when the other is a 'real illusion' and the 'real illusion' becomes what is 'real' when the 'real' becomes the 'real illusion'.

The 'real' and the 'real illusion' go back and forth depending upon where it is one is located when one examines one or the other, as one experiences one or the other, as one finds oneself immersed 'within' one or the other.

Once 'within' the concrete, it is the reality of abstraction which takes on the appearance of being an illusion but which in fact is a 'real illusion', a 'functional' illusion,

Once 'within' the abstract, it is the reality of the concrete which takes on the appearance of being an illusion but which in fact is a 'real illusion', a 'functional' illusion.

The real and the 'real illusion' do not become unimportant to the whole for the whole cannot exist without the two for the two are integrated as one.

Each is dependent upon the other. The one, the Cartesian, the concrete, is the engine of the other, the non-Cartesian, the abstract.

And the other, the non-Cartesian, the abstract is the 'creator' of the other, is the 'creator' of its own engine, is the 'creator' of its means to 'grow' as opposed to stagnating or decaying away.

What do the boundaries of zero dimensions, the means of separation, have to do with the concepts of abstraction and the physical? The boundaries are what they are.

What are these black circles?

The black circles are just what they were said to be.

The black circles are, in essence, nothingness.

The black circles are 'something', they are nothing and as something have functionality as nothingness.

The black circles, nothingnesses, are the means by which unique individuality expresses itself. T

he circles are a means of individuality retaining its individuality through the existence of the somethingness of nothingness itself.

With the black circles of 'nothingness' one obtains an understanding of individuality, one obtains an understanding of unique and distinct individuality:

Positive Rational Numbers = {#'s > 0 which can be expressed as a/b where a and b are whole numbers and where b ≠ 0}

Tunnel of Positive Rational Numbers:

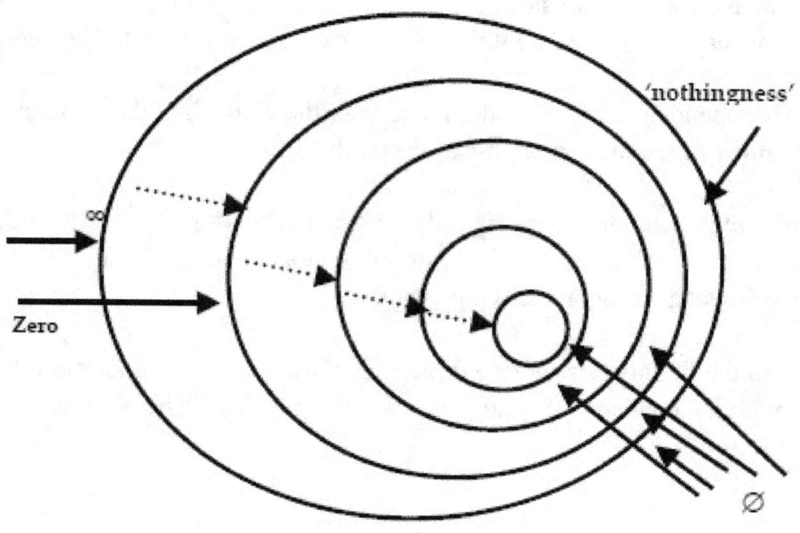

Without the separation of individuality created by the boundaries of 'nothingness', one loses order, distinctiveness, individuality and the diagram becomes:

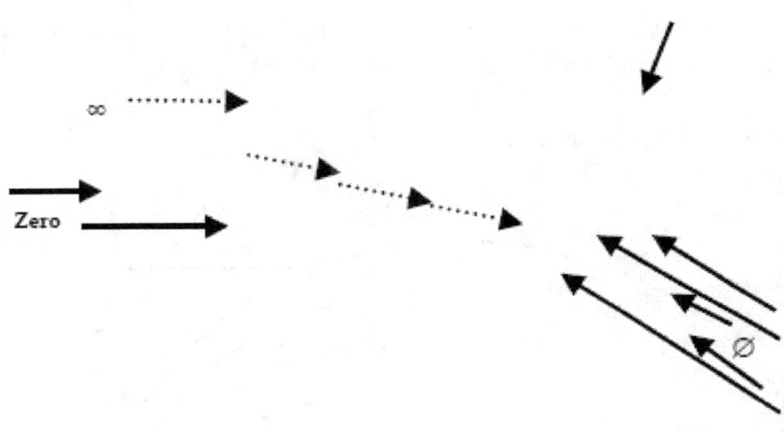

So? So chaos reigns and becomes seemingly meaningless but as we shall see as we proceed, what seems to appear 'to be' need not 'be' what is.

This paradox of chaos and lack of order evolving out of order is simply what all paradoxes are. This paradox is simply a lack of perceptual understanding and as such the paradox disappears with the infusion of perceptual understanding, abstract understanding.

Ok, ok, once again, now what?

Now we return to our task of understanding abstractual functionality through expanding upon our understanding of numbers, which is but one of many infinities to the infinite power forms of abstractual potentiality.

Positive Rational Numbers = {# > 0 which can be expressed as a/b where a and b are Whole numbers and where b ≠ 0}

Tunnel of Positive Rational numbers:

Now becomes:

Rational Numbers = {#'s which can be expressed as a/b where b ≠ 0 and where a and b are integers (as opposed to being Whole numbers)}

Tunnel of Rational Numbers:

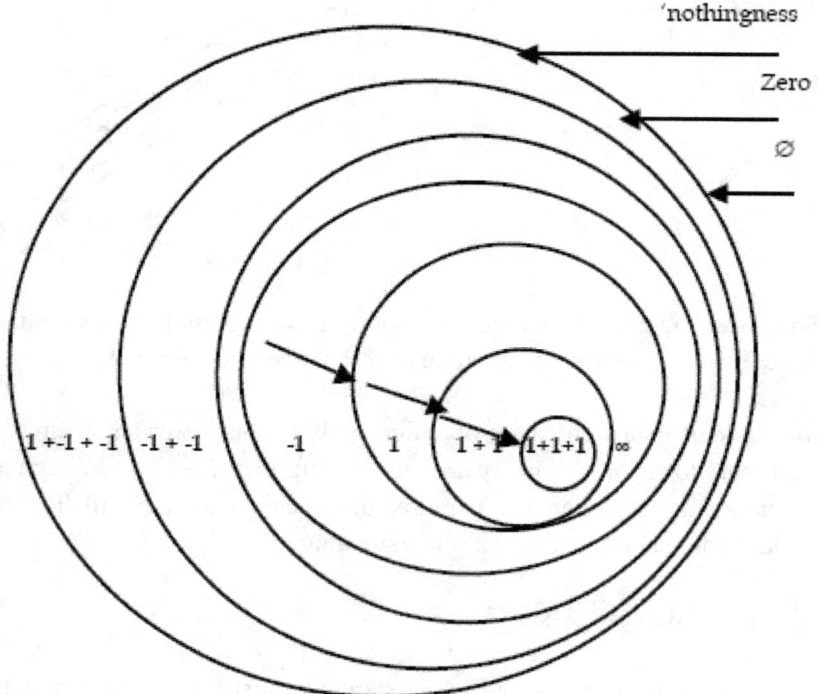

The tunnel is gaining greater depth as we look down the tunnel from our position within the tunnel.

It appears to be gaining the dimension of depth through the process of 'moving' in one 'direction' so lets gain a better understanding of this 'expanding dimension' by turning around and looking in the other 'direction'.

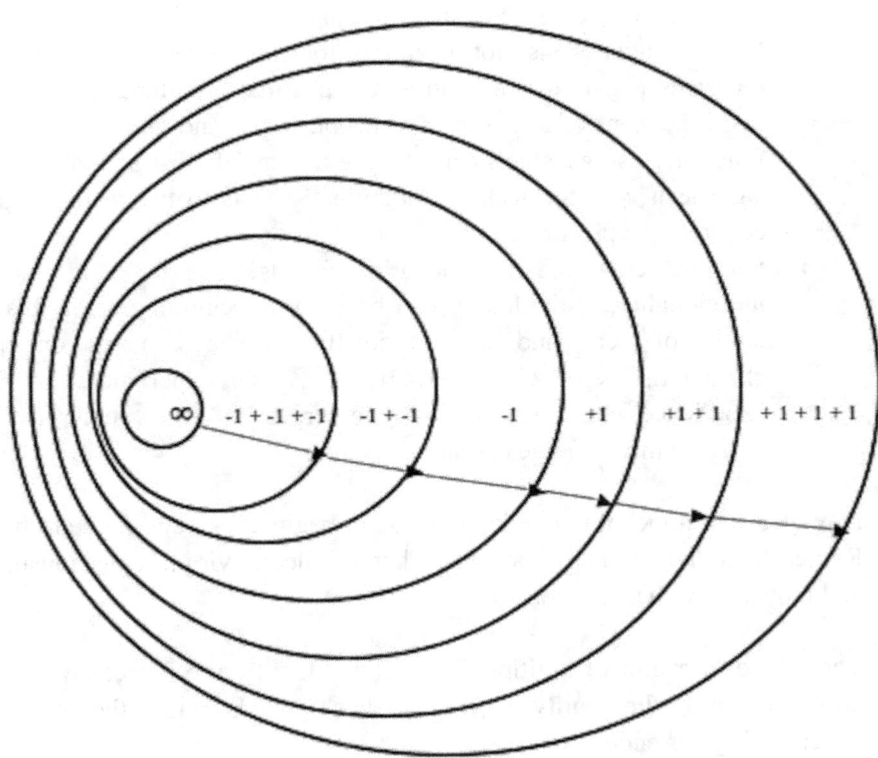

Now several characteristics emerge:

---

1. The tunnel retains its division of individuality through the use of 'nothingness'.
2. Another aspect emerges which demonstrates the need for our previous apparent digression regarding the concept of Zeno and nothingness, which is that nothingness:
    a. takes up no space
    b. does not exist within the reality of the concrete

c. does exist 'within' the realm of the abstract
d. is the only 'thing', which retains its same characteristics 'within' both realities, 'within' the reality of the concrete, and 'within' the reality of the abstract.
e. is, within both the reality of the concrete/physical and the reality of the abstract, 'nothing'.

3. It, 'nothingness' is not negation for as we see in the above diagrams negation is as real as the positive. 'Nothingness' is not just a concept of a lack of a concept, it 'is' and as such it has a function just as significant to the realm of abstraction as the significance of the lack of 'nothingness' has to the realm of the concrete/the physical.
4. 'Nothingness' finds NO location for itself as its own unique individuality. Individuality can be seen as beginning at the black circle of zero and as individuality expands, takes on the characteristics of ½ its experience, ¾ its experience, 7/8 its experience etc. until it finally emerges as 'one', emerges as individuality, emerges as an individual.

Now you will notice the individual does not begin at one and move to two. Rather the individual begins at zero, what is called a 'virgin' consciousness and moves on to become 'one'.

The whole of a unit of multiplicity of individuality never becomes 'one' however until the entity's journey has ended, until the entity's experiencing has ended.

The second individual, two, does not necessarily begin after one has ended its journey, rather the second 'one' begins at zero, at its virgin consciousness, just as the first 'one' began at zero.

In terms of individuality, our graphic now becomes:

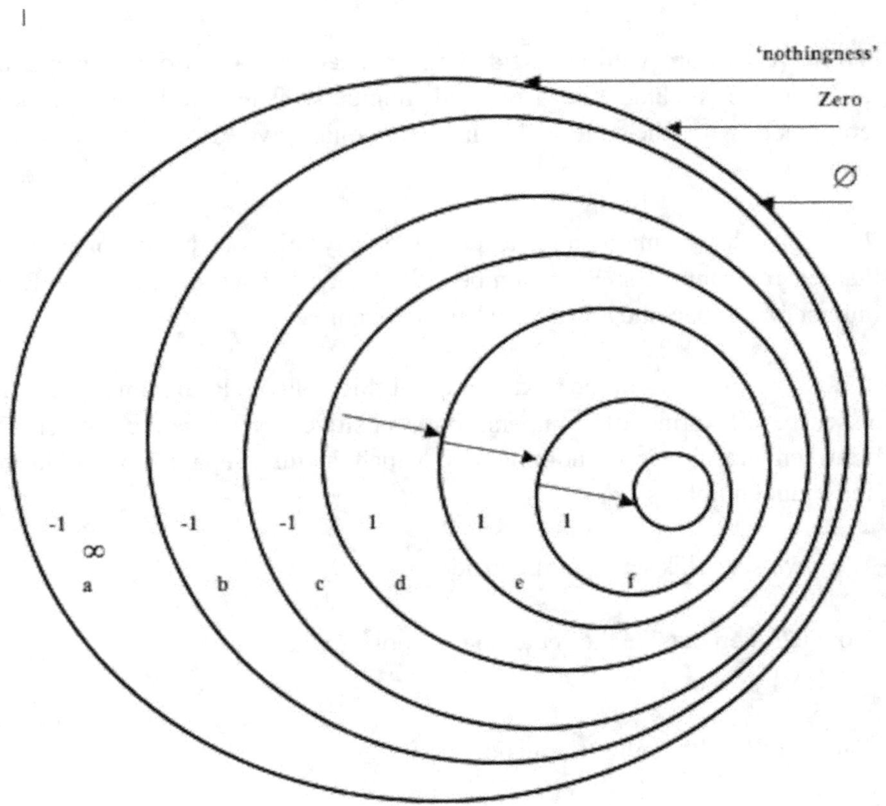

In this diagram the totality of the number, the totality of the individual number, the totality of individuality is composed of all its experiencing.

Thus 1/2 becomes a part of what makes up one and 1 1/2 becomes a part of what makes up the second individual.

Zero, 'Nothingness' remains just that nothingness. The characteristic of nothingness exists as itself whether nothingness exists in the realm of the negative numbers, the positive numbers, abstraction, or the concrete.

To move forward with understanding we need to expand our concepts of numbers to include the Irrational numbers. The Irrational numbers combined with the Rational numbers would give us the set of Real numbers.

Ironically, it was mathematicians not metaphysicians or philosophers who named real numbers: Real numbers. The significance of the name Real numbers will become apparent when we examine

Imaginary numbers in the next section of this volume. Real numbers are, in essence, all forms of numbers, both positive and negative, which fit between the circles of nothingness, which in turn separate our units of individual numbers.

We have seen this set from the inside.

Now lets step outside the set and get a look at what we have:

Tunnel of Real Numbers – outside view:

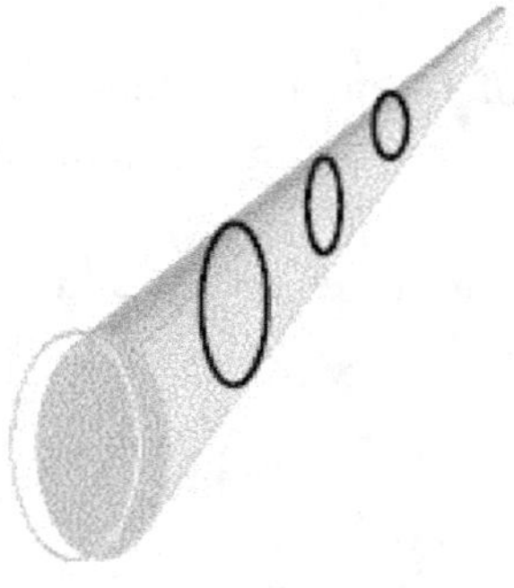

Keep in mind; the circles divide individual number from each other.

These partitions are actually zero, nothing, yet they allow the uniqueness of each number to exist independent one form the other.

Each number in turn is simply one unique number beyond the next and each number emanates from the point of its virgin origination and proceeds to its point of termination of experience which for numbers moves from the point zero to the point of being a whole unit. For numbers, what falls between these two points, zero and wholeness, are the concepts of Rational and Irrational.

It is interesting that mathematicians have named these sets independent of metaphysical thought. Surely, one has to wonder about the concept of 'pure coincidence', serendipity.

Not only do the names, Rational, Irrational, Real, Imaginary, etc have huge implications for metaphysical thought but so to do the concepts the names metaphysically imply in terms of the new metaphysical system of the individual acting within God.

The simultaneous correlations seem to be too 'purely coincidental' to be ignored. Perhaps this in itself is a testimony regarding the application of Ockham's Razor 'across' subject matter boundaries.

If we expand the 'length of the tunnel of numbers we obtain what appears to be an endless division-less tunnel. We 'know' however that the tunnel is composed of unique elements of individuality of numbers. As such the 'endless' division-less tunnel is not what exists, rather what exists is an endless tunnel comprised of unique units of individuality.

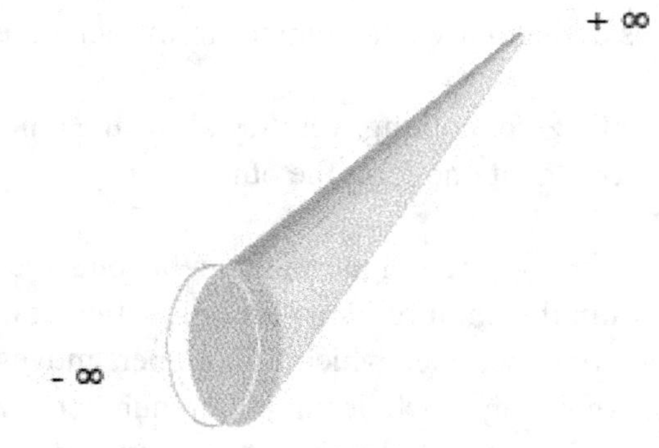

6. Imaginary Numbers

We have reached the outer limits of the Real numbers. Mathematicians at this point discovered they have a problem. There are mathematical operations we cannot perform with the existence of this set alone.

The square root of the number one can be found within the tunnel. The square root of one is plus or minus one since: $+1 \times +1 = 1$ and $-1 \times -1 = 1$.

One cannot, however, find the square root of a -1within the tunnel. In order to establish a solution to the square root of a -1, one must create a new set of numbers.

Having established such a set, one must then name the new set. Appropriately enough, mathematicians, after having established the new set, named the new set of numbers: Imaginary numbers.

The solution to the problem, the square root of a -1, now becomes +/- i. 'i' represents what we call the imaginary number of... It is the task of mathematicians to determine if this set has a one to one correspondence to the set of Real Numbers.

The task of metaphysics is to determine how the concept of imaginary numbers relates to metaphysical systems and in particular to the new metaphysical system of the individual acting within God.

The question for metaphysicians regarding the individual acting within God becomes: How does the concept of imaginary numbers, imaginary things, fit into a system of non-Cartesianism powered by Cartesianism, the individual acting within God?

It is graphics and in particular the graphics we have already established that will help us understand the answer to the stated question.

With a simplistic understanding of imaginary numbers in place we now find our tunnel of numbers looking more like the following:

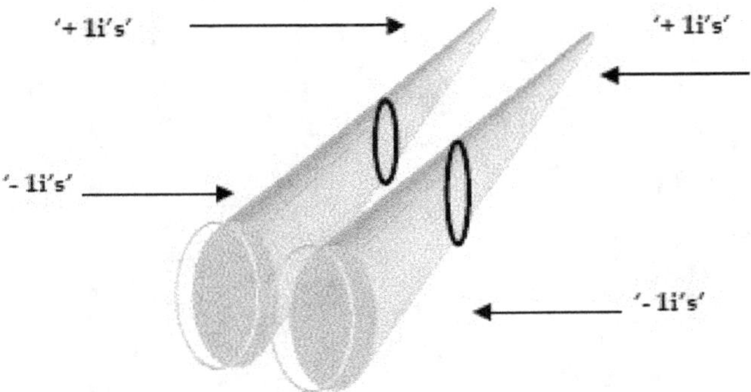

We now have two tunnels of numbers. Keep in mind that numbers are but one abstract concept existing within the realm of abstraction. The tunnel of numbers are composed of numbers which give an appearance of repelling each other since they line up in a straight line form. Applying the principle of repulsion to the two tunnels of numbers we would get a configuration similar to the structure of an organic molecule of methane – one carbon atom and four hydrogen atoms. In our present analogy carbon would be zero and the four sets ('+1's', '-1's', '+1i's', '-1i's') would be hydrogen. Chemist or physicist will have no difficulty picturing the configuration. The rest of us will have little choice but to proceed as best we can.

This formation begins to take on the following resemblance:

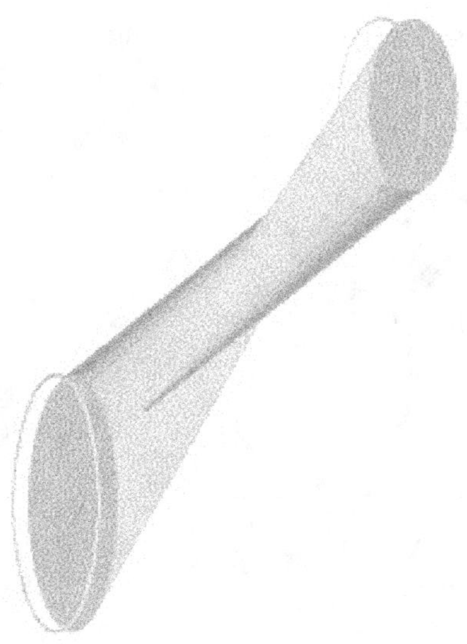

The two sets appear to pass through each other.

In terms of the two dimensional plane within which we are working, the two sets appear to be at right angles to each other.

Mathematicians with an organic chemistry background can begin to think of this as the existence of four sets of hyperbolas representing the set of numbers which are not members represented by sets ('+1's', '-1's', '+1i's', '-1i's') symmetrically placed within the realm of three dimensions. Inside each of the hyperbolic sets of numbers, numbers which are not members of themselves would be found, wrapped within 'nothingness', the sets of number which are members of themselves represented once again by the symbolic sets ('+1's', '-1's', '+1i's', '-1i's').

So it is simplicity begins to yield to complexity within the metaphysical system of the individual acting within God.

To simplify the complexity, lets simply illustrate the diagram as follows:

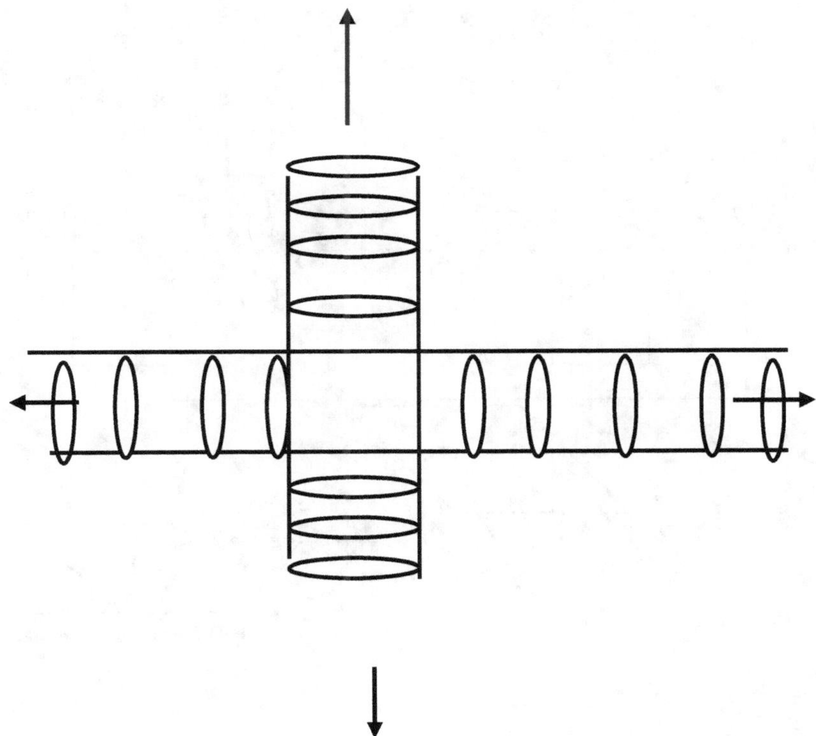

The diagram presents a problem. The two tunnels intersect at the zero point but they do not do so in a clean fashion. So lets insert a concept into our tunnels.

Let's insert time into the tunnels.

As we can see from the second diagram of this volume, time is what we call the fourth dimension; therefore as we insert time into the graphic, we will simultaneously insert the three dimensions of length, height, and depth represented by the first diagram of this chapter.

Appling the two actions give us the following tunnels that intersect at zero.

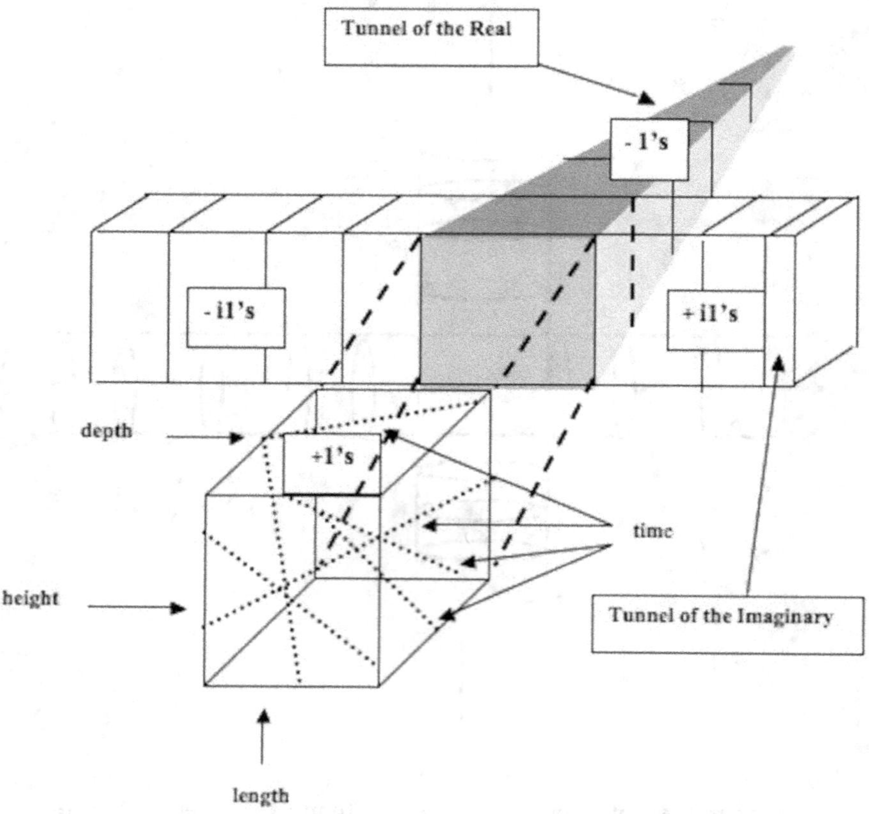

The tunnels are now square prisms through which time flows. Time flows in both the tunnel of the Real and the tunnel of the Imaginary.

At his point, we are going to stop referring to what the mathematicians call the 'Imaginary' and start referring to the concept of the 'Imaginary' as a 'real illusion'.

Now we have the Real and the 'real illusion'. In addition we will stop referring to the Real as the 'Real' and refer to 'Real' as the 'real'.

This may appear to be insignificant but the upper case 'R' implies the Real is more real than the 'real illusion' and within the realm of the abstract no one abstraction is more real than another.

We now have two sets of numbers represented by tunnels incorporating time.

It is critical we not lose track of the concept that the tunnels intersect at zero, have their individual unit numbers begin at zero, begin at nothing. From this location of nothingness the individual unit numbers develop.

Each individual, each individual unit number, individuality/multiplicity develops from this lack of 'all', develops from the lack of all experience which in turn is represented by the complexity immerging from the set of all numbers found between the origination of the number as an individual number and proceeding to and through the full development of the number being a total unit.

The development of the complete number brings the graphic up to and through the point of virgin beginning from which the next number, unique individual not only begins but also includes as a part of itself.

The graphic gains a more stylistics appearance if one reduces the graphic to:

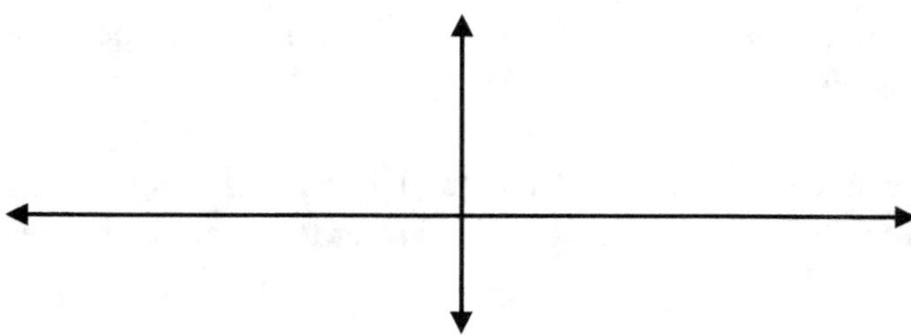

However, we are not attempting to understand stylistic methods but rather we are attempting to conceptually understand metaphysical ideas such as individuality, virgin beginnings, real, real illusions, time within…, time throughout…, through the study of mathematical symbolism.

Ironically, the next step in understanding the concept of individuality of numbers is to erase individuality itself through the process of eliminating the bonding element of zero found within the tunnel of the real and the tunnel of the real illusion. Once having erased all the partitions of zero, we obtain what might be termed: The incoherency of individuality within which time is immersed

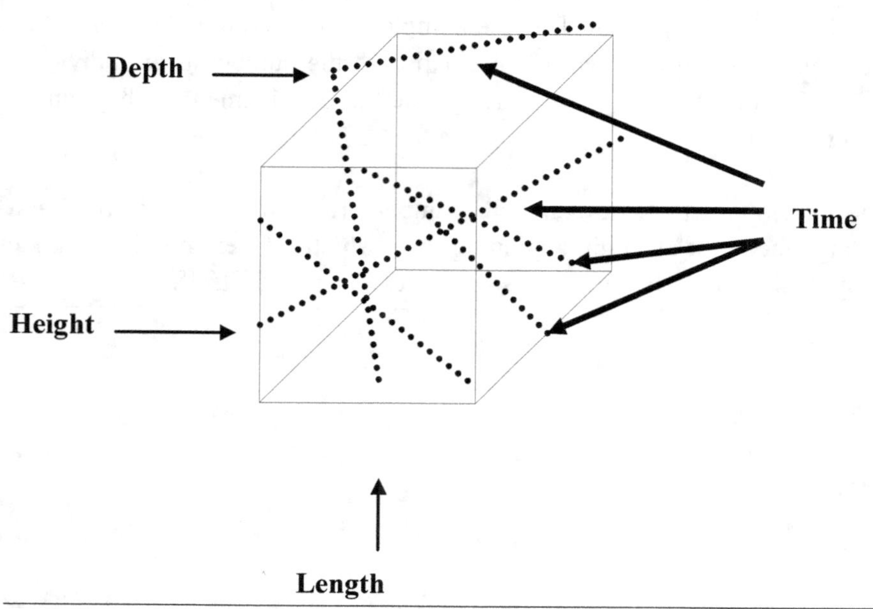

Look familiar? Now we are going to reproduce this four times and name them.

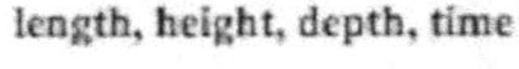

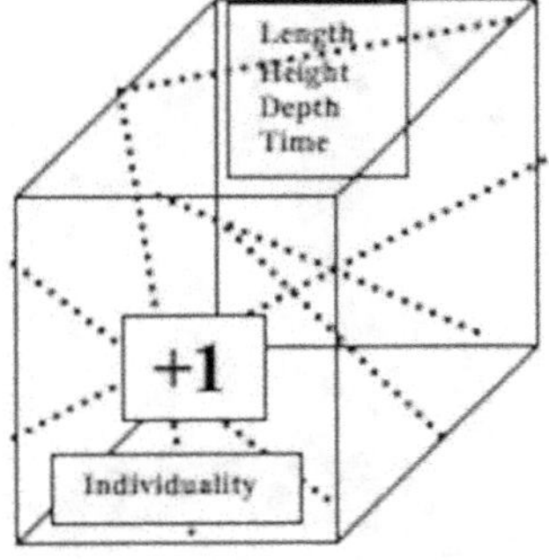

+ 1

length, height, depth, time **Name**

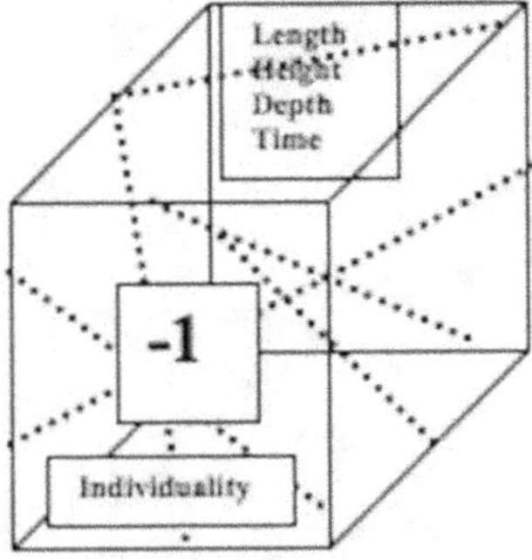

- 1

length, height, depth, time **Name**

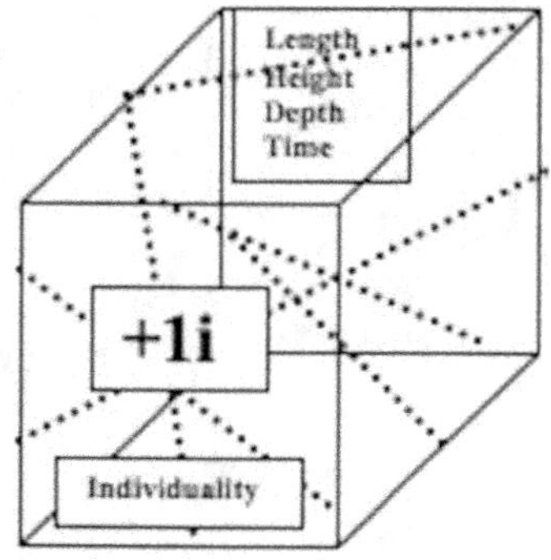

+ 1i

length, height, depth, time **Name**

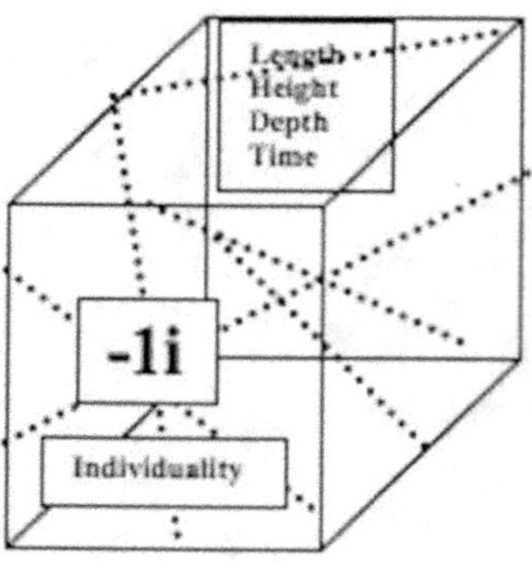

- 1i

If we think of each tunnel as a fusion of infinite numbers of unique individual numbers and then surgically cut them apart at exactly the boundaries which separate them from each other, which of course is of zero thickness and represented by a partition of nothingness itself we obtain:

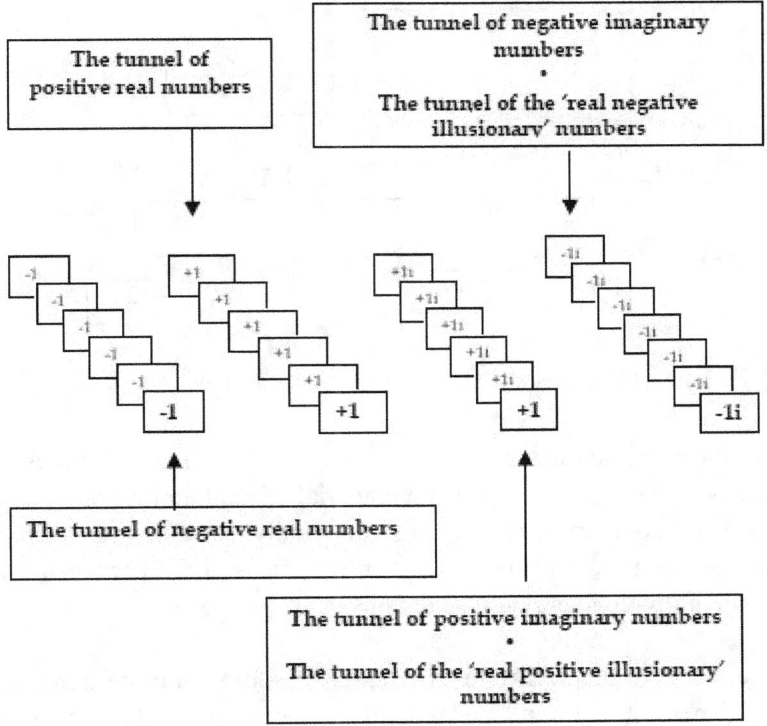

Which in turn begin randomly floating throughout space-less-ness taking segments of time with them, taking segments of individuality of time, taking segments of what one might label pieces of unique individuality.

We then have randomness of time and individuality.

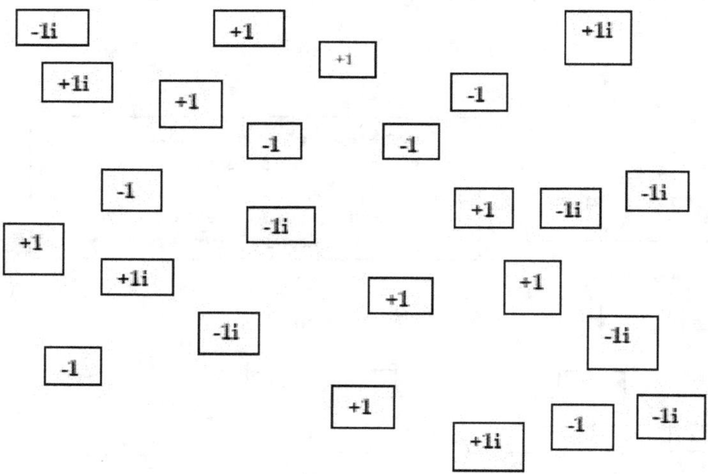

We then have randomness of time found within the randomness of individuality as well as 'room' for the potential of random motion, random movement, random experiencing of the individual in the lack of multidimensional time and the lack of multidimensional distance located 'outside' the individual and outside the physical.

In a sense we have the ability to experience an almost infinite number of experiential permutations independent of time other than the time found 'within the individual itself.

The graphic thus becomes:

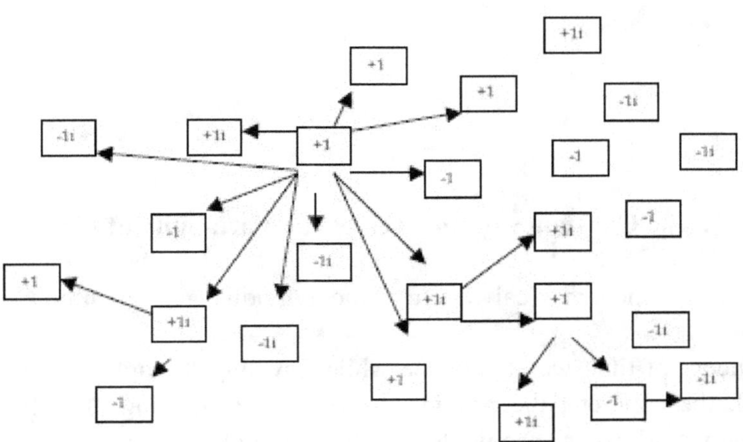

The degree of motion, the variety of experiential permutations becomes increasingly large as the number of unique pieces of individuality increase.

It would appear, as individuality increases to infinity, the number of experiential permutations would take an infinite amount of time to experience all the possible sequential permutations possible.

Remember, however, that time is found within the pieces of individuality not 'outside' the pieces of individuality and as such the experiencing of each possible permutation takes no time but rather becomes an experience of time itself occurring within timelessness itself.

7. The Constancy of time verses the Variability of time

In terms of the metaphysical system of the individual acting within God:

The concept of time we experience while traveling through what we call the real, the concrete, the physical universe is something different than what one would experience while traveling through the real illusion of the abstract, the region 'outside' the sequential cause and effect aspect of time.

The concept of individuality encapsulating time capable of experiencing within what we call the real illusion, experiencing what we call the abstract, experiencing what lies 'outside' the boundaries of our universe would experience void the sequential cause and effect limitations time and space place upon 'things' immersed 'within' time and space.

Sequential time exists in the physical universe and due to its very nature of being sequential has a sense of constancy.

This constancy is not one of relative constancy from one location to another, or relative constancy 'between' objects, but rather a constancy of sequentiality.

Within the universe, for any particular piece of awareness, there is a constancy of beginning at zero awareness; there is a constancy of beginning at virgin consciousness and ending at the summation, totality of awareness in terms of that unique individual.

On the other hand, within the realm of the abstract, the constancy of time can only be found 'within' the unique piece of individuality 'floating about'. As such time is not relative to the unit of individuality /multiplicity found 'within' the purity of abstraction since time is not a universal fabric of existence of the abstract.

Pieces of individuality found within the purity of abstraction can therefore experience all forms of experiential permutations capable of being composed from the infinite number of multiplicity found 'within' the abstract.

Once all permutations of experiences for a unique entity of individuality is complete, then all has been experienced for that packet of individuality in terms of that packet's experiencing what physicality has to offer regarding 'what was', 'what is', and 'what will be'.

Since experiencing 'all' takes place within the purity of abstraction, how does boredom generated by Nietzsche's 'eternal recurrence' of action through experience become circumvented for abstractual existence and how is uniqueness of individuality preserved? Wouldn't each unique individual simply 'become' the same as another?

The only way presently conceivable for abstraction to circumvent 'eternal recurrence' and homogeneity is for the totality of abstraction to develop, create, and implement a means to continue generating new, unique pieces of individuality capable of perceiving uniquely based upon unique perceptual outlooks in turn based upon unique experiencing which uniquely perceive.

But wouldn't the very process of developing, creating, and implementing take time and thus wouldn't time be a component of abstraction itself as opposed to being locked within individuality found within abstraction?

By definition time and space would be a universal fabric of abstractual totality itself since time and space are factors of mass and energy (E = mc(2), E = m (d squared/ t squared) and thus 'belong' by mathematical definition to the 'universal fabric' of the concrete, physical reality.

Thus limited one-dimensional time sequentiality is only a function of the individual's perception and only found within the beam of time known to us as a form of our unique universe, the physical universe.

Wouldn't the above paragraph imply the potential exists regarding the existence of 'other' universes existing based upon other universal fabrics other than time.

That is exactly the type of creative thinking which is capable of emerging as a form of scientific speculation initiated by the new metaphysical perception of non-Cartesianism powered by Cartesian, the individual acting within God.

That, amongst a host of other implications, is exactly why this metaphysical perception propels metaphysic once again into the forefront of both science and religion where it belongs.

In terms of the metaphysical model of the individual acting within God, how is time, found within physical reality, assembled in an orderly fashion, assembled in a coherent fashion, as opposed to existing in the form of incoherency of time found within abstraction located 'outside' the physical?

The key to assembling time in a coherent fashion may lie in the concept known as light amplification by stimulated emissions of radiation. The concept of light amplification by stimulated emissions of radiation is generically known as 'laser'.

The 'laser' tool takes incoherent packets of energy and restructures their relative random cohabitation into a form of coherent reinforcing orderliness. In short order emerges from chaos.

Replace ‘l’ with‘t’ and one gets the concept of taser: time amplification by stimulated emissions of radiation.

The radiation in this case is a form of incoherent time, a form of incoherent packets of time, beginning-end sequentiality within which abstraction itself can experience, undergo the sequence of virgin consciousness to completion of the unit of individuality.

What then becomes of this newly formed unit of individuality?

The unit of unique experiential individuality adds to the summation of the totality of endless packets of unique experiential individuality and increases the resultant possibilities of permutations available to both the whole/singularity of abstraction/knowing and individuality/multiplicity of abstraction/knowing in an exponential manner.

In essence one obtains a non-Cartesian system of knowing/knowings powered by a Cartesian system generating ‘newness/growth’ of new knowing, one obtains the individual acting within God, one obtains panentheism.

Is this then what we mean by the concept of Einstein and ‘i’? Actually no it isn’t.

The previous discussions ‘mean’ nothing, rather what the examination of all this ‘stuff’ did was lay the groundwork for understanding both the concept of Einstein and ‘i’ as well as understanding the implications the concepts Einstein and ‘i’ generate for science, religion, and philosophy through the field of metaphysics..

Before we move into the concepts of Einstein and ‘i’, we will need to understand the concepts of Newton and ‘i’. However, before we examine the concepts of Newton and ‘i’ it will help to summarize the progress we have made through a graphic format.

In essence we have been exploring the difference between the constancy of time and the variability of time. The constancy of time vs. the variability of time can be demonstrated as follows:

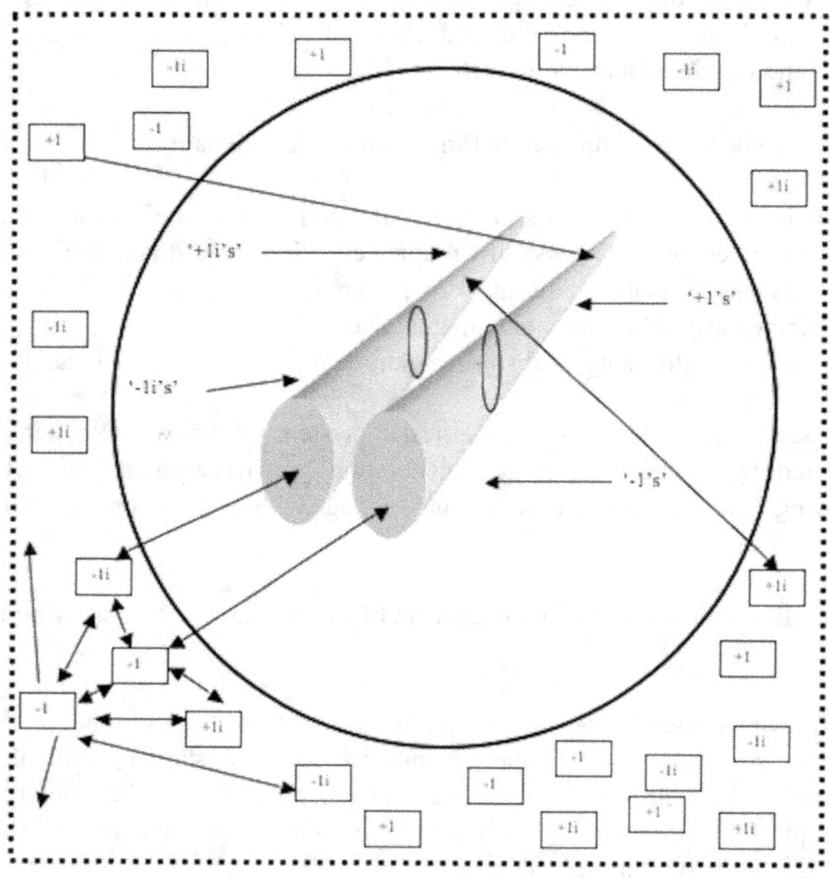

Part II: Resolving the issue with a new metaphysical perception

Part IIa: The Newtonian 'i' - Velocity Equals Distance Divided by Time

---

8. Introduction

It is not the historical development of Hegel that is to be addressed in part II of this volume but rather we will examine Hegel in terms of Newtonian physics and the concept of the inverse.

This is the second introduction to be found within this volume and again it must be emphasized that I do not pretend to be a mathematician nor do I pretend to be a physicist. The concepts put forward within this volume are completely suggestive in nature but need to be said. The mathematics is not what is important here but rather it is the direction to which the suggested mathematics points that is of importance.

The purpose of this volume is to stimulate mathematicians and theoretical scientists towards an idea regarding the concept of what 'total' reality might be versus our present day concept of having no idea regarding what 'total' reality is.

This section deals with Newtonian physics in the attempt to lead us to Einstein and Einstein's concept of relativity as it relates to metaphysics. Before Einstein can be examined, two concepts must be addressed:

1. A metaphysical understanding regarding the linear relationship which exists between time and distance and the relationship which

exists between inverse time and inverse distance must be established.

2. A metaphysical understanding must be established bridging Hegel and Einstein. The bridge to be built is a metaphysical understanding rationalizing the linear relationship which exists between the direct proportionality of time and distance using Newtonian physics. Once having built the bridge using the linear concepts of Newtonian physics we will be able to explore the significance of quadratic concepts suggested by Einsteinian physics. The mathematics of Newton and Einstein led to the confirmation of both Aristotelian Cartesianism and Kant/Hegelian non-Cartesianism. The results of this confrontation lead to a new metaphysical system incorporating both Aristotelian and Kant/Hegelian concepts

Where do we begin this tedious process of understanding the primitive relationship of Newtonian metaphysics and Einsteinian metaphysics? We begin with Hegel.

9. Expanding knowing revisited

History does not direct us, as Hegel suggested, for to be 'directed' and then to have no choice but to follow is simply determinism. for 'To follow' implies the path has already been established..

We direct history through our actions of free will. History is simply a directional vector of human action and as with all vectors, the direction the vector points can change with a change of any one of the two components creating the vector.

To change the direction of a history vector requires energy input as does 'changing' any vector. The energy required to change the history vector must originate from the element from creating the history vector.

The questions then become: What is the element/vertex from which the vector itself emanates and what are the two components comprising the history vector of humanity?

Whether one is a determinist or a believer in free will, the answer is the same: The element/vertex from which the history vector emanates is the entity of knowing, the entity capable of awareness of the history vector itself and capable of intentionally applying energy to the vector in an attempt to change the history vector's direction.

The history vector can be described as a combination of actions generated by the passive state of 'perception' and the active state of action generated by the entity of knowing.

The operative term here is not 'action' nor is the operative term 'perception'. The operative term is 'generates'. We thus obtain a history vector emerging from an identified source and driving through time through the action the source initiates.

The process is a three step process:

1. Perception generates action.
2. Actions generate reactions.
3. Reactions generate social ambience.

The result of steps one through three is the creation of a history vector which may quiver as it moves through time but which does not, cannot, change its resultant direction unless the source of the vector itself changes its perception of itself.

The two components creating the history vector of humankind involve two operative words:

1. The first operative word in the three step process is explicit in nature.

The first operative word is 'perception'.

To 'change' the direction of the vector one must change perception.

To leave perception as it is and to change action may cause change but it is only temporary change for the perception remains as it was and eventually actions will revert to the natural function of their natural emergence from perception.

The result: Without changing perception, the long-term change of human action remains ineffectual.

2. The second operative word in the three step process is implicit in nature.

The second operative word is ‘change’. If the product of perception is not what one chooses to change then one should not ‘change’ perception.

If the history vector of humankind is pointed straight and true towards its mark, is pointed in the direction one wishes humanity would continue, then one should declare the death of metaphysics and be done with it.

Strangely enough, ‘action’, singular, is not one of the operative words of this three step process for action ‘follows’ perception.

‘Reactions’, plural, are not one of the operative words for ‘reactions’ follows ‘action’. And finally, ‘social ambience’ is not one of the operative words for it is neither a verb nor is it a base component because ‘social ambience’ emerges from the concept of vast numbers of ‘reactions’ within society.

Philosophy, with the acceptance of Hegel’s metaphysical system being ‘the’ ultimate of metaphysical systems, declared the death of metaphysics and as such has attempted to be done with metaphysics other than giving metaphysics tokenistic acknowledgement.

But to be done with metaphysics has been a more difficult task than originally expected. Although philosophers may wish to be done with metaphysics, humanity subconsciously relates to the concept of meta – beyond physics – the physical.

Humanity has not embraced the existence of the history vector which appears so overtly human. Humanity has not accepted the violent and aggressive desire to dominate and subjugate which humanity finds coursing rampantly through its veins.

As such it is humanity, which has refused to ‘let go’ of metaphysics and it is humanity, which refuses to let the majority of the intellectuals rule what is to be the direction towards which the vector of human action points.

Humanity remains idealistically hopeful regarding the principle direction its actions will eventually take.

Humanity senses its true nature lies in compassion, tolerance, and pluralism versus its present day primitive state of harsh exclusionism and aggressive desires to dominate.

Humanity senses the conflict existing between these two states remains a conflict because we lack an understanding, a perception, of the whole of reality and humanity senses that it is the understanding of the whole of reality which will change our perception of the whole of reality being simply the physical.

In short humanity is doing its best to protect its sense of its true nature, its sense that it, humanity, is deep-down by nature compassionate, tolerant, pluralistic, and inclusive.

Humanity has shown great resolve in preserving the idealistic perception it has of itself.

Humanity intuitively senses the significance of individuality/multiplicity being 'the' ultimate of meaning for itself as opposed to society/singularity and until someone comes along and rationally explains why society/singularity is not the true nature of humanity, the history vector of humankind will continue to precariously point towards individuality/multiplicity being the course humanity is to travel. The history vector appears to be evenly split between the concept of the individual and that of society being the ultimate significance.

The vector is susceptible to a shift causing it to point more significantly towards one or the other. Humanity is waiting, eagerly, and at times despairingly for a rationalization of its, humanity's, importance.

With the significance regarding the inner human turmoil having been verbalized we can now begin examining the forces composing the history vector of humanity:

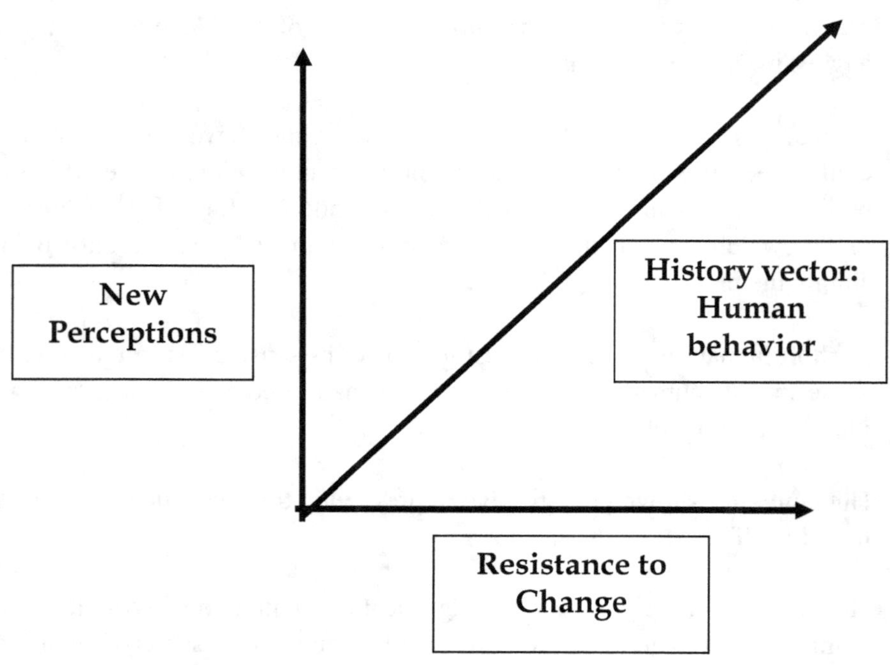

If resistance to change is high and development of new perceptions is low the vector becomes the 'x' axis. If resistance to change is low and development of new perceptions is high the vector becomes the 'y' axis.

In both instances constancy of consistency becomes the base to social behavior. The constancy is either complete predictability or completely unpredictability.

Hegel sensed we must accept what it is we learn as we move throughout history as opposed to discarding all and beginning anew.

> *'His (Hegel's) aim is rather to construct a new system, a new philosophy, a theory which will include all that is positive, every conceptual advance.'*
>
> **<u>Before and After Hegel</u>**, Tom Rockmore, p. 2, 1993.

If we are to follow the metaphysical path Hegel suggested, it would appear we must find a metaphysical system incorporating Aristotle through Kant – Cartesianism as well as Hegel through the present – non-Cartesianism.

- In short, to accomplish Hegel's goals, we must incorporate the two fundamental metaphysical systems which presently exist both in the sense of their simultaneously existing yet in the sense of their existing independently.
- This leads us to two possible choices 'separation through exclusion' or 'separation through inclusion'.
- The issue regarding the metaphysical 'separation through exclusion' versus 'separation through inclusion' is fully addressed in Volume 13: Russell and Volume 13: Metaphysical Systems.

Volumes one through seven addressed paradoxes put forward by past great metaphysical thinkers.

The question becomes: Is it possible to incorporate the intuitions of Zeno, Aristotle, Boethius, Copernicus, Leibniz, Kant, and Hegel into one system? Examining our present metaphysical perceptions the answer appears to be: We presently have no metaphysical perception, which incorporates all such past philosophical intuitions and the presently existing metaphysical systems which incorporate the thoughts of these great thinkers themselves are riddled with major philosophical paradoxes.

The reason the present day fundamental metaphysical systems find themselves riddled with major philosophical paradoxes is that the presently existing metaphysical options each profess to be 'the' system and as such exclude the competing systems as being invalid systems.

In short, the present day fundamental systems which compete for the status of being 'the' ultimate metaphysical system, all operate within the framework of 'separation through exclusion' one from the other as opposed to 'separation through inclusion' of all.

All our present day fundamental metaphysical models reject some aspect/s of the others.

In short, metaphysically we as a species are ignoring Hegel's 1st principle: Do not exclude what we have learned from the past but rather build upon what it is we have learned.

Resolutions to the paradoxes which riddle the presently recognized metaphysical models would then appear to lie in a different approach than 'separation through exclusion'.

If we incorporate a process of separation through inclusion we find it is possible to build a metaphysical model capable of incorporating past intuitions of great thinkers without the simultaneous development of major philosophical paradoxes.

Such a model was examined in terms of resolving the paradoxes created by such metaphysical thinkers as Zeno, Aristotle, Boethius, Copernicus, Leibniz, Kant, and Hegel in found in the work: The War and Peace of a New Metaphysical Perception.

Through the process of examining the previous seven metaphysical thinkers it becomes apparent that Hegel reopens the system, which Aristotle closed and thus Hegel brings us full circle, bring us back once again to Zeno:

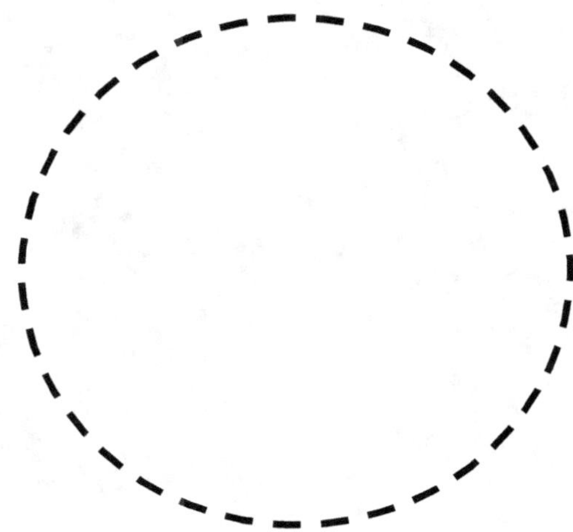

Granted there is now ‘more’ ‘inside’ the system but the system remains as Zeno saw it – open.

So what lies beyond the physical, beyond (meta -) physical (- physics)?

10. The constant (k) variable

10 a. The 'constant' factor of variability

The 'variable' constant, the 'constant consistency of change', or the 'constant variable', they are all the same. The three phrases describe the same perceptions from different points of view.

One is the tail of the elephant, one is the ear of the elephant, and the other is the trunk of the elephant as 'viewed' by the same blind man.

All three concepts:

1. Variable constant
2. Constant consistency of change:
3. Constant variable:

are located within the same region of existence:

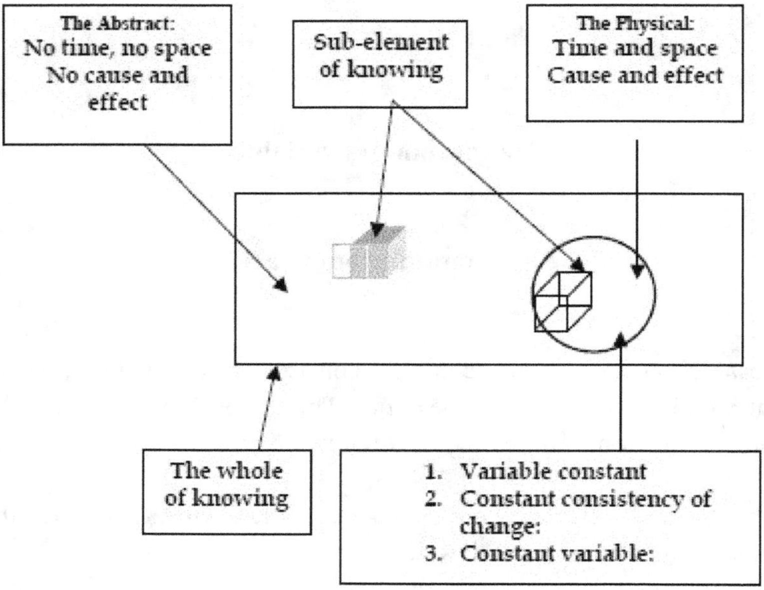

The Abstract:
No time, no space
No cause and effect

Sub-element of knowing

The Physical:
Time and space
Cause and effect

The whole of knowing

1. Variable constant
2. Constant consistency of change:
3. Constant variable:

Strangely enough it is 'within' the realm of the physical where the three aspects of change occur, where variability occurs.

The region 'beyond' the physical is void time and space as a universal fabric.

The region 'beyond' the physical is where time and space are found within the sub-elements of existence as opposed to being a universal fabric within which the sub-elements of awareness are immersed.

It is within the physical where variables are to be found and were change evolves.

It is 'outside' the physical where the permutations, and combinations of events, as represented by multiple sub-elements of knowing as well as by the singularity of the whole itself, is expressed by the factorial of the sub-elements plus one.

What then of the analogy regarding the perception based upon a blind man and his relative position to the whole of perception?

1. Variable constant (k)
2. Constant consistency of change (k):
3. Constant variable (k):

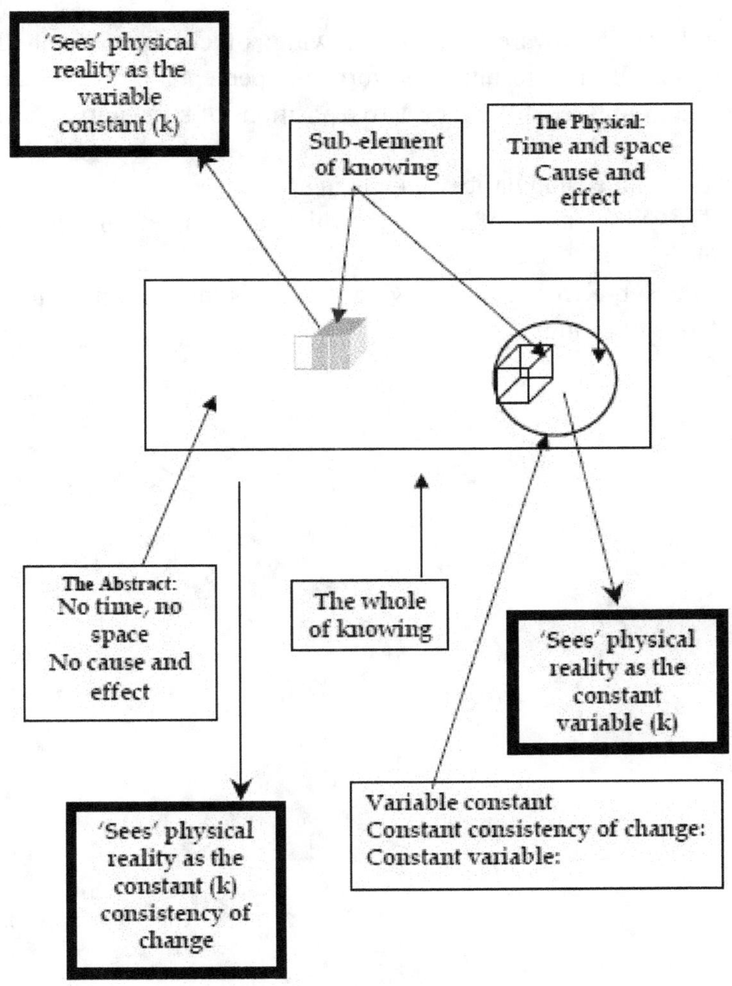

What then of physical reality? What then of the 'region' of time and space 'within' which a sub-element of knowing fills itself with knowing of time and space?

The constant (k) of change is found in three 'locations' for the concept of change is only significant once it is 'known', is only known once awareness of the change is perceived.

As such one finds the awareness, the knowing, of change only in the knowing entity itself. The result: three forms of perceptions of the same concept emerge from three different relative positions of knowing:

1. The whole, singularity, of knowing
2. The sub-element of knowing which is complete in terms of change
3. The sub-element of knowing which is metamorphosing in terms of change

10. The constant (k) variable

10b. The 'constant' variable of physicality

Part I. Hegel introduces the first mirror: Inverse physicality

V = k: The constant variable equals physicality

If we begin our examination of Hegel from the perspective of the time within which Hegel existed, we begin to realize we must examine the interrelationship of the physical and the abstract in terms of Newtonian physics for Einsteinian physics had not yet been developed.

$$v = \frac{d}{t}$$

where:

v = velocity
d = distance
t = time

Three concepts are involved with motion/action:

1. Velocity
2. Distance
3. Time

In terms of metaphysics, however, only two concepts are involved:

1. The physical
2. The abstract

Again we come back to Zeno's concepts regarding seamlessness and multiplicity from which the paradoxes of motion emerged.

There is, metaphysically, a third element involved.

The third element is not physical nor is the third element abstractual in nature, rather the third element is a line of demarcation.

The third element, the line of demarcation is the equal sign found within an equation.

In metaphysics the equal sign becomes a mirror where the left is the left and the right is the right however in relationship to the image opposite itself the left for the object on one side of the mirror is the right for the object on the other side of the mirror.

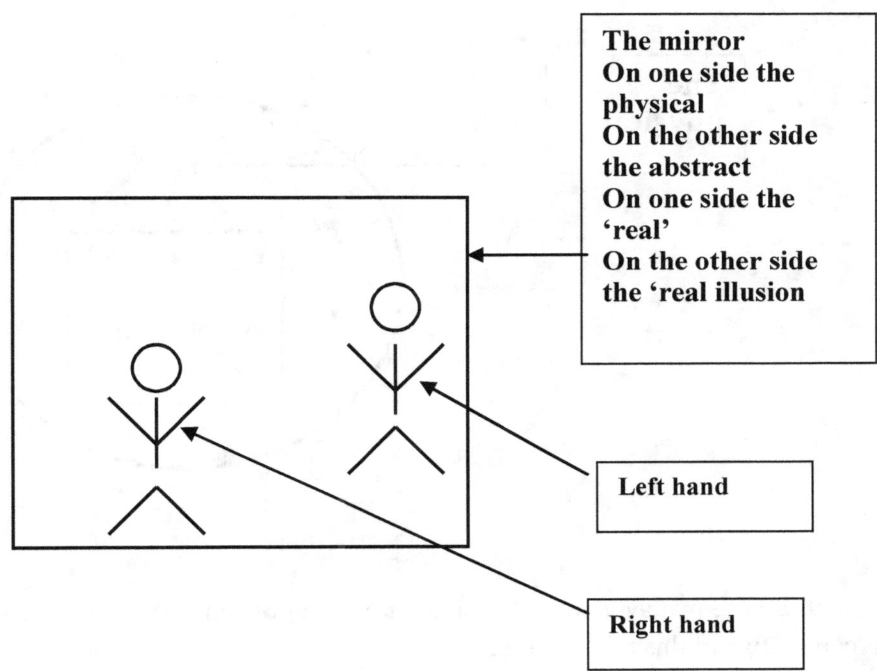

If we examine the concept of the relationship of the abstract within the physical we see that 't' is inversely proportional to 'v' and 'd' is directly proportional to 'v'.

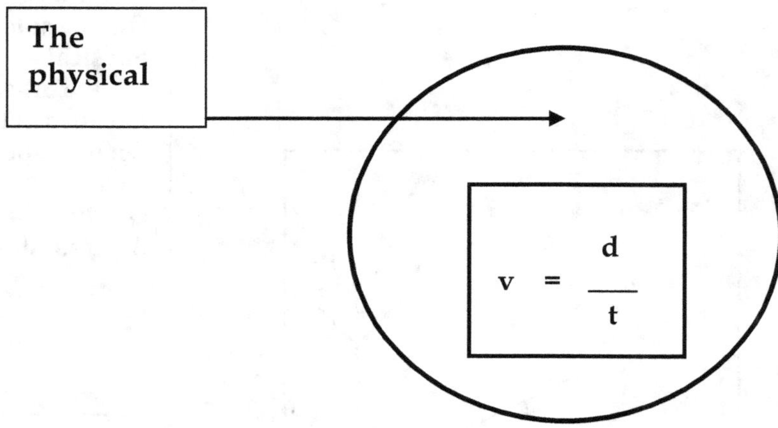

With a little mathematical manipulation we can obtain a slightly different perspective of this relationship.

d

v =

t

becomes

$$vt = d$$

Metaphysically ‘v’ becomes what might best be termed a ‘constant of physicality’.

The constant of physicality in essence refers to the fact that the number changes as a function of abstraction found within the physical and thus ‘v’ is a variable within the physical but the number does not change as a function of abstraction found within the purity of abstraction and thus ‘v’ is the constant ‘k’ within the abstract.

It may help to understand the previous complex statement through the use of graphics.

In the use of graphics we will assign a more common mathematical letter to ‘v’ in order to designate a ‘constant of physicality’ which when examined in terms of a region of physicality, the constant varies but when examined in terms of a region of pure abstraction, the variable of velocity does not vary since in the region of pure abstraction, time and distance do not exist as the fabric of universality and thus ‘v’ becomes a constant, becomes the ‘constant of physicality’.

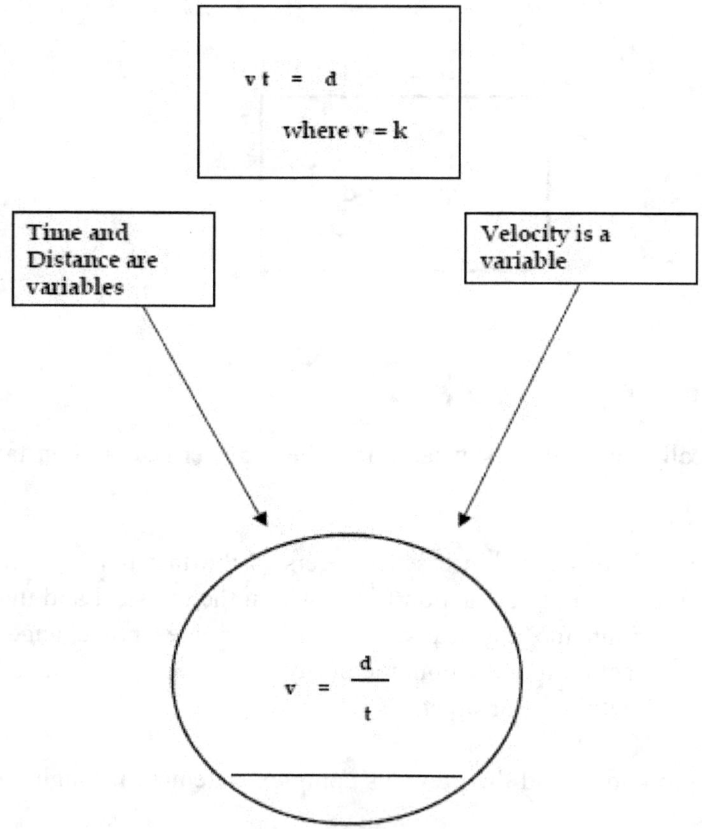
v t = d
where v = k
Time and Distance are variables
Velocity is a variable
v = d / t

How is it that 'v' is a variable yet 'v' is a constant?

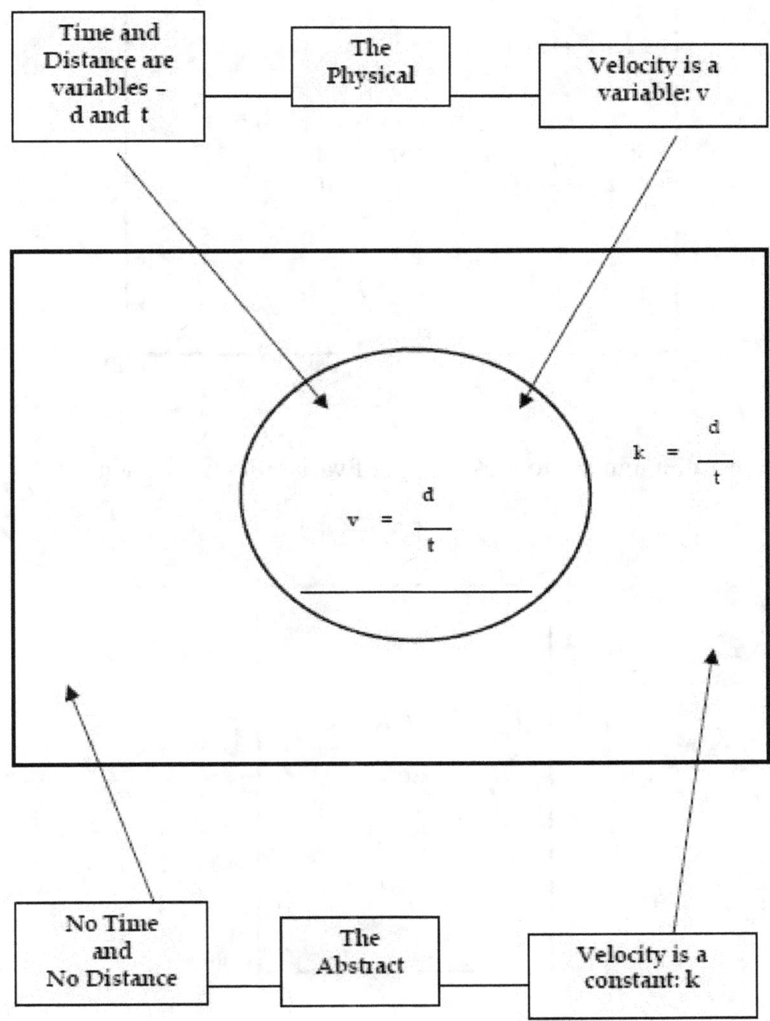

As such we obtain, in terms of abstraction:

k t = d

where: k = a constant of physicality

t = 0

d = 0

If we assign a unit number to the concept of variability, we obtain k = 1

k t = d

where k = 1

then:

1 t = d

t = d

Now the question becomes: Where are time and distance the same entity? In the physical, time and distance are not the same entity.

But in the purity of abstractual existence, time and distance are the same entity for both time and distance/space are entities, which do not exist as universal fabrics within the purity of abstraction.

In the purity of abstractual existence, existence is does not exist immersed 'within' time and 'within' distance/space and as such neither time nor distance/space exists but rather time and distance/space exist as perceptions found 'within' individual entities of knowing/the individual/multiplicity and as such zero and infinity represent both time and distance/space simultaneously.

A graphic will help us understand such a statement:

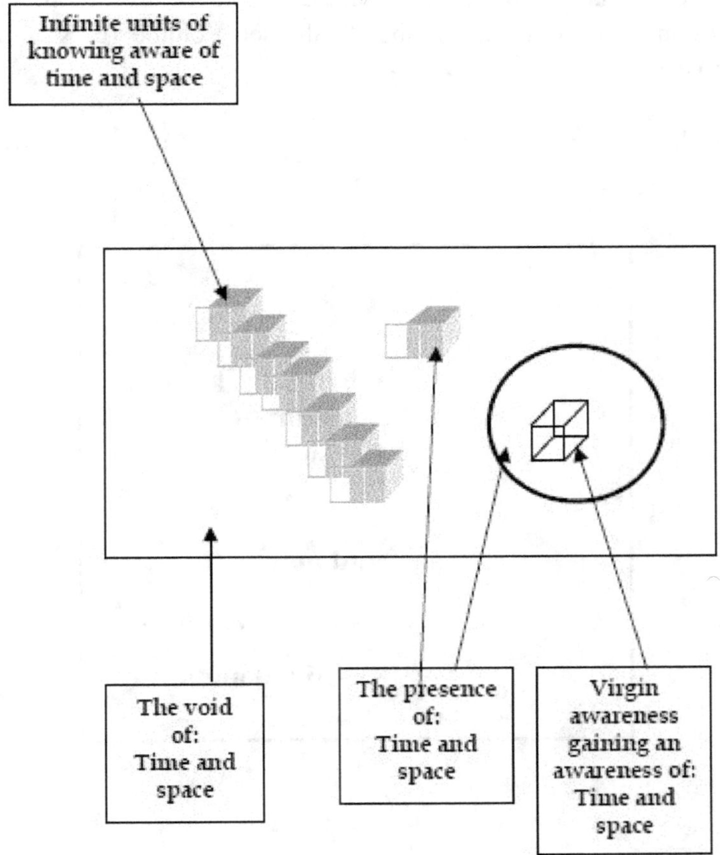

Within the region void time and distance/space, time and distance/space are nothing and all things can be accomplished without regard to time and distance/space limitations.

The limits placed upon knowing found within pure abstraction is the limit placed by 'what is'.

'What could be' places no limits upon knowing found within pure abstraction for 'what could be' 'is not' nor does 'what could be' have the potential to be.

The detailed examination of the last complex sentence was thoroughly presented within previous volumes (specifically see Volume 10: Kant and Volume 11: Hegel).

Thus we now have:

$$kt = d$$

where: k = a constant of physicality

and now:

$$t = 0 \text{ and/or } \infty$$

$$d = 0 \text{ and/or } \infty$$

Time equals zero within the realm of pure abstraction. Time equals infinity within the realm of the physical.

Distance/space equals zero within the realm of pure abstraction. Distance/space equals infinity within the realm of the physical.

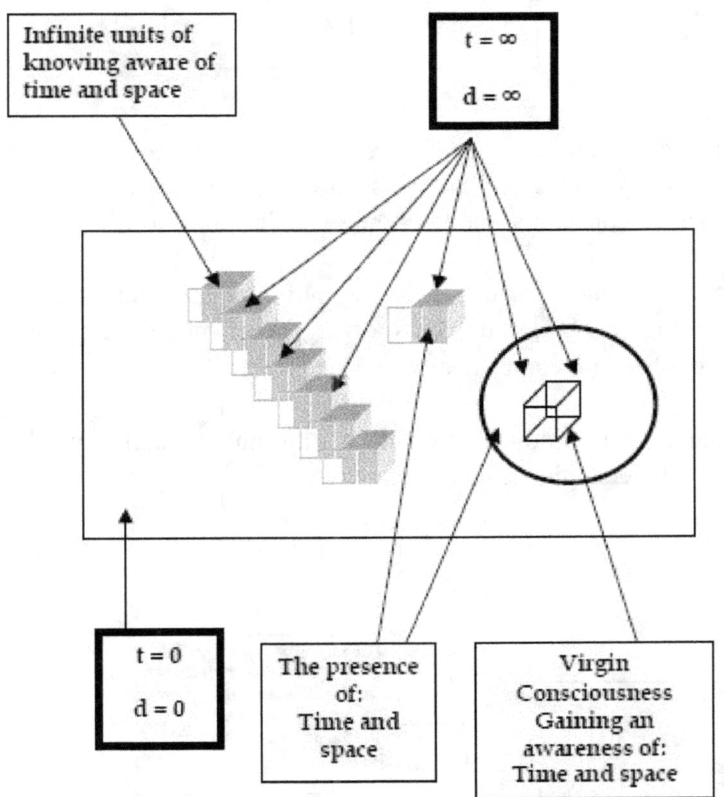

If v is the constant of physicality: 1 or 'k'

Then it follows that if v is one and v' or the inverse of v is also one and thus:

$$v / v' = d / t$$

and

$$v' / v = t / d$$

The metaphysical understanding regarding the given concept is best expressed as a mirror image of the physical constancy of change or what could be called the constant variable or the variable constant 'k'.

The concept of change found within the physical but experienced by abstraction through multiple units of knowing experiencing uniquely was perhaps best described metaphysically by Hegel.

Thus it is that we will credit Hegel with having unknowingly introduced the first mirror: Inverse physicality

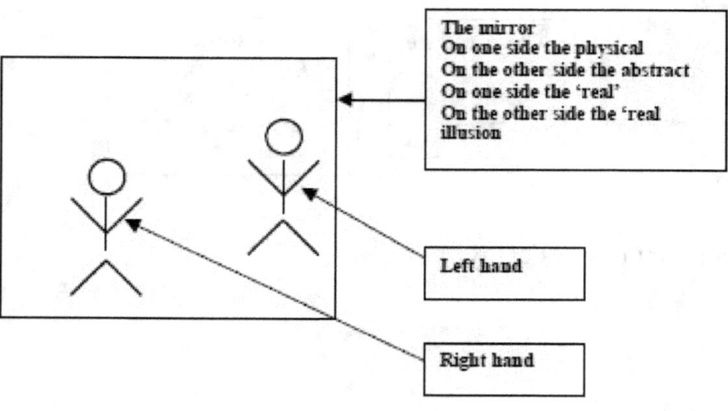

10. The constant (k) variable

10b. The 'constant' variable of physicality

Part II. Einstein introduces the second mirror: The 'i' inversion

One mirror is insufficient to explain the concept of Einsteinian physics as it relates to metaphysics.

Rather than develop the second mirror as we did Hegel's first mirror, we will start by introducing the second mirror into the graphics and then proceed backwards to Einstein's famous equation: E = mc(2) at which point we will proceed to examine the significance of the second mirror to a new leap in metaphysical perceptions.

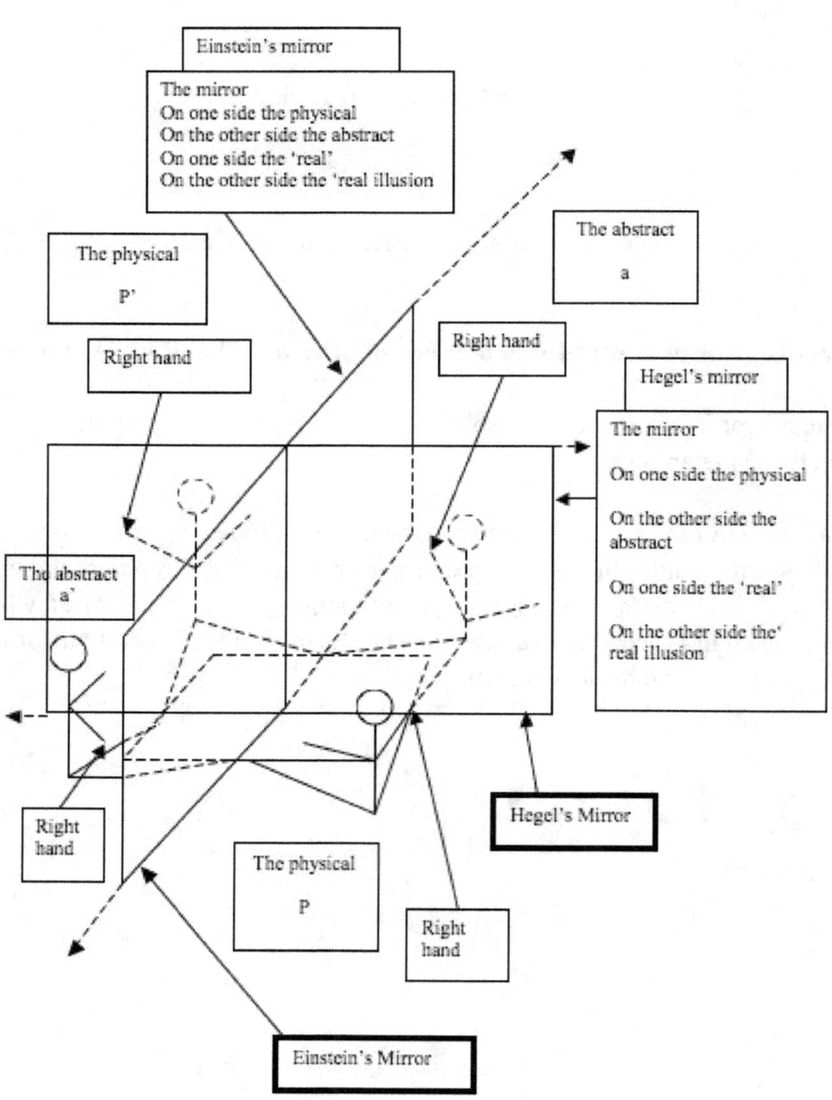

We can now see we have two regions of physicality, p and p', and we have two regions of abstraction, a and a'. From such a perception we can begin to understand the concept of two regions of velocity, v and v'.

11. d = t

$v = d/t$	$v' = 1/v$
$vt = d$	$1/v = t/d$
$v = k$	$v = k$
$kt = d$	$1d/k = t$
$k = 1^*$	$k = 1^*$
$1t = d$	$1d/1 = t$
$t = d$	$d = t$
or	or
$t + -d = 0$	$d + -t = 0$
or	or
$0 = d + -t$	$0 = t + -d$

For time minus distance to be zero and for distance minus time to be zero, distance and time need to either be the equivalent of each other and existing in the same form since the equations demonstrate their difference to be equivalent to zero itself or both time and distance need to be nothingness itself in terms of our understanding what it is we perceive to exist within our physical reality.

Since we do not perceive of distance and time as being equivalent forms of existence, we will assume the two, distance and time are non-elements of the physical in terms of either being physical in nature.

What nature could we then assume the two, time and distance to be? The two, distance and time, could be abstractual concepts void the characteristics of physical ness.

Since neither would exist as a universal fabric of the abstract, the question becomes: How is it that v and v' can both exist, metaphysically speaking, in the form:

$d + -t = 0$
$0 = t + -d$
$t + -d = 0$
$0 = d + -t$

$t + -d = 0$
$0 = d + -t$
$d + -t = 0$
$0 = t + -d$

How or where is it possible for such a relationship to exist in a fashion where each maintains their own unique identity?

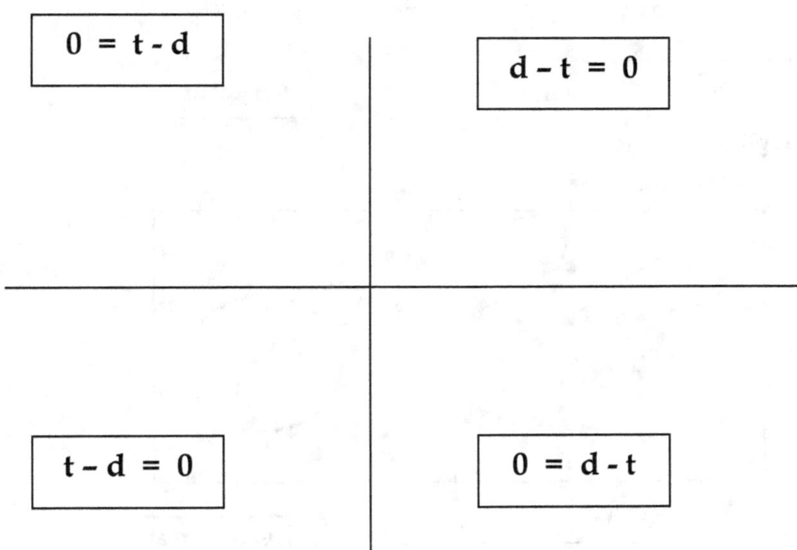

The two, distance/space and time, could be abstractual concepts void the characteristics of physical ness. Time and distance/space being abstractual could exist as a form of abstraction, which would not exist as a universal fabric of the abstract.

Graphically such a concept could be demonstrated as:

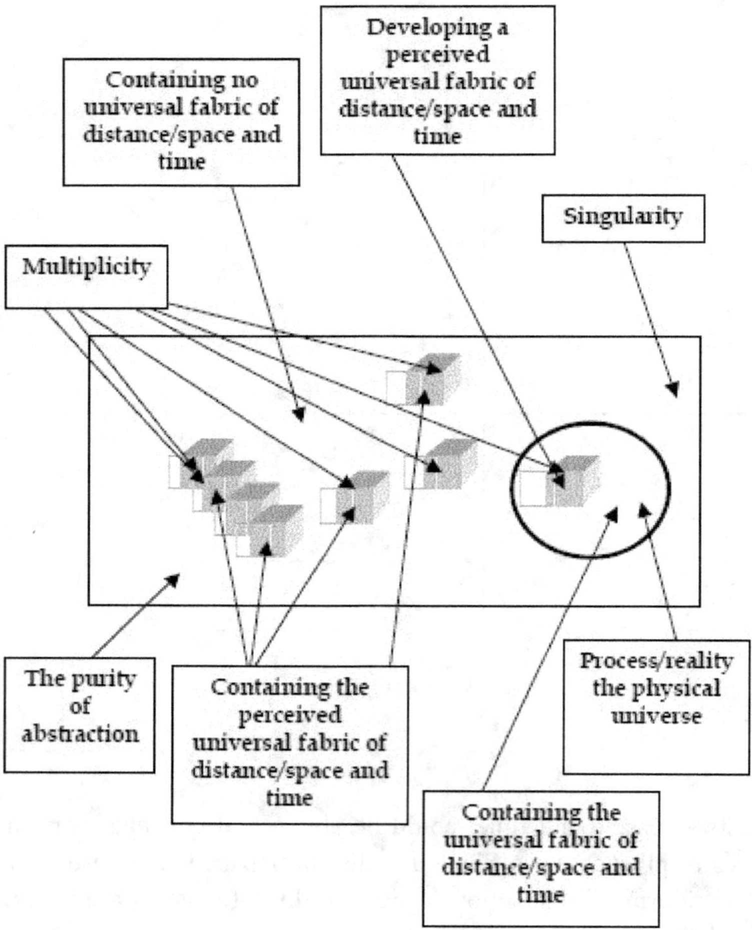

12. 1

The whole is the whole:

100 / 100	=	1
100 / 10	=	10
100 / 1	=	100
100 / .1	=	1000
100 / .01	=	10000
…		
…		
100 / 1/∞	=	∞
…		
100 / 0	=	∞

The whole divided by itself is one. One becomes understandable in the metaphysical sense as being the whole.

Metaphysically wholeness, 1, exists in two states:

1. the wholeness of the whole in terms of what lies 'outside' the whole
2. the wholeness of the whole in terms of what lies 'inside' the whole

With the metaphysical system of Aristotelian Cartesianism and the metaphysical system of Kant/Hegelian non-Cartesianism, we obtain two different perspectives of the wholeness:

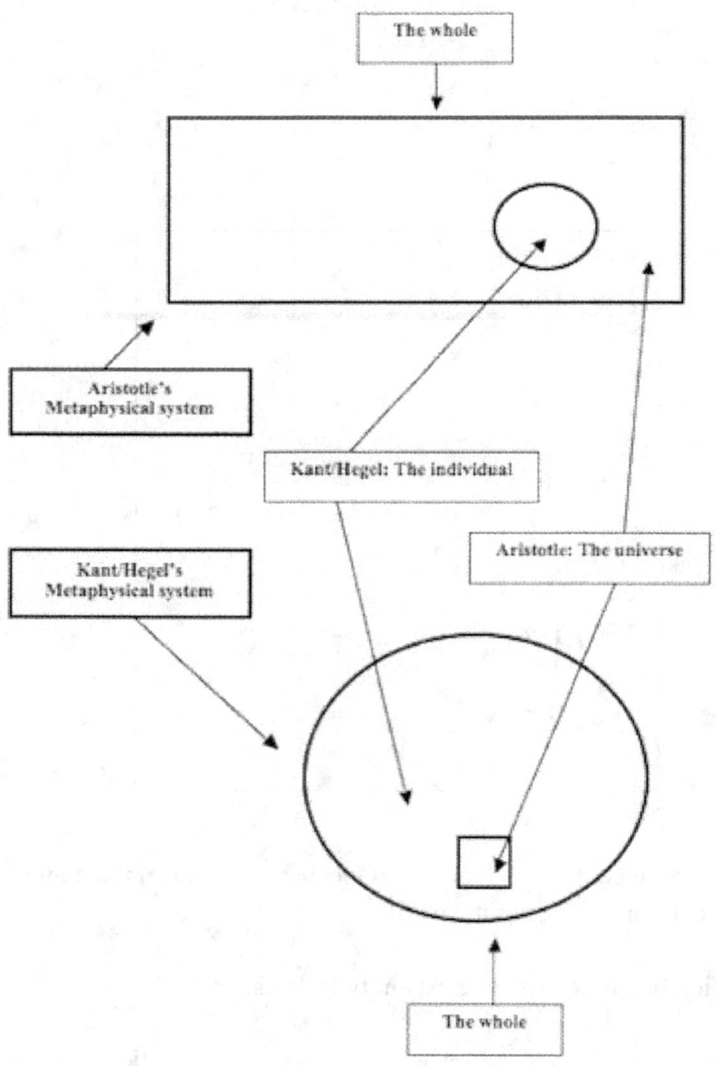

In the Aristotelian system, the whole is found outside the individual and is composed of the multiplicity of individualities found outside the individual and includes the individual – multiplicity abounds.

Mathematically such a concept is expressed as a ratio which in turn can mathematically be expressed as the operation of division, the separation of the whole into its unique entities of existence:

$$\text{Fraction} = \frac{\text{The whole / one}}{\text{The total parts of the whole}}$$

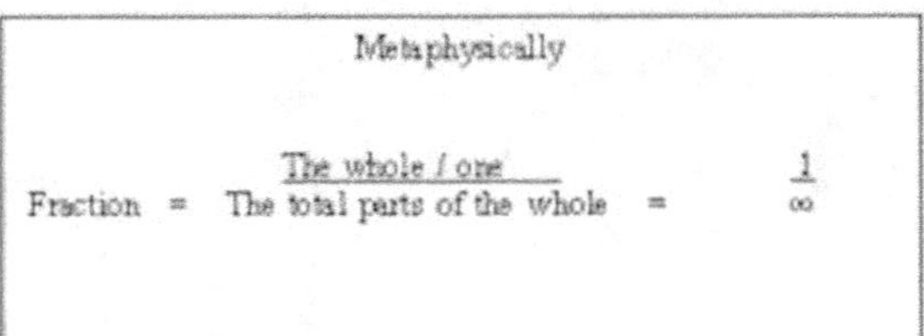

Metaphysically
Fraction = The whole / one
The total parts of the whole = 1
∞

Graphically the concept can be demonstrated as:

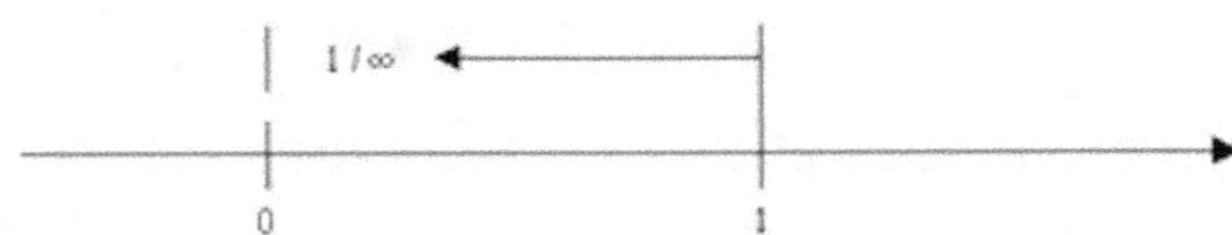

We obtain the whole composed of its infinite parts.

This is not the only means by which we can demonstrate the whole composed of its infinite parts.

The whole composed of its infinite parts can also be demonstrated as:

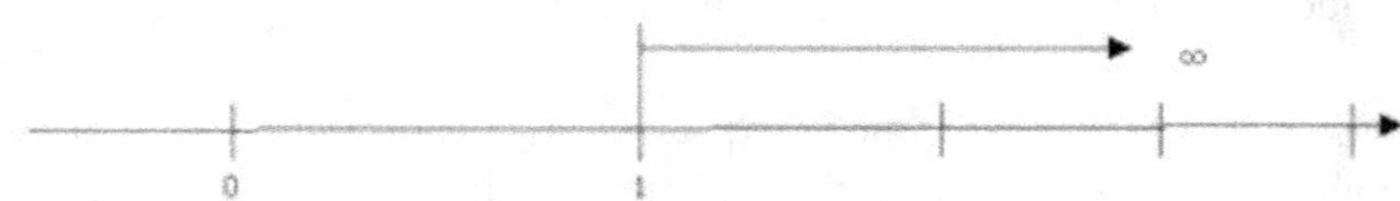

Giving us:

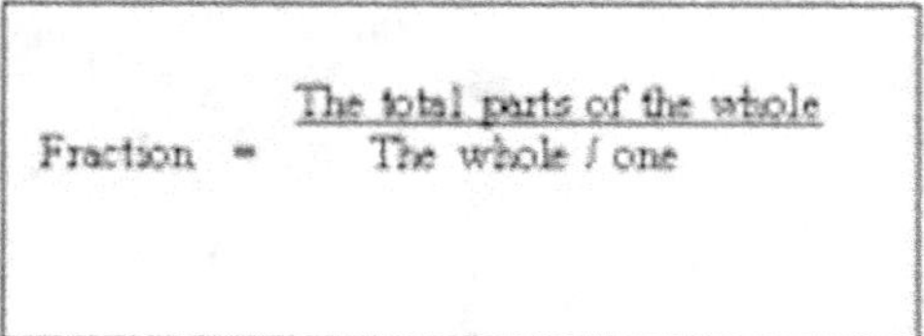

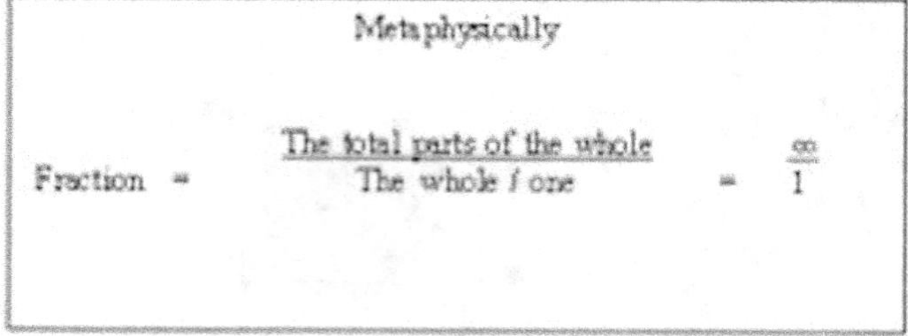

In both scenarios the number one/the whole/singularity is significant, just as the concept of infinite parts/individuality/multiplicity is significant.

In essence we have:

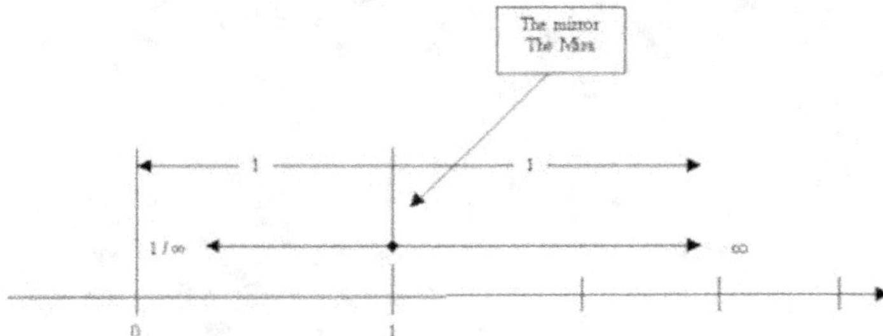

The metaphysical mirror is not the point zero but rather the metaphysical mirror is the point represented by the unit number one/the whole/singularity.

Mirrors invert the image. Just as the mirror causes the movement of the right hand to be the movement of the left hand of the image.

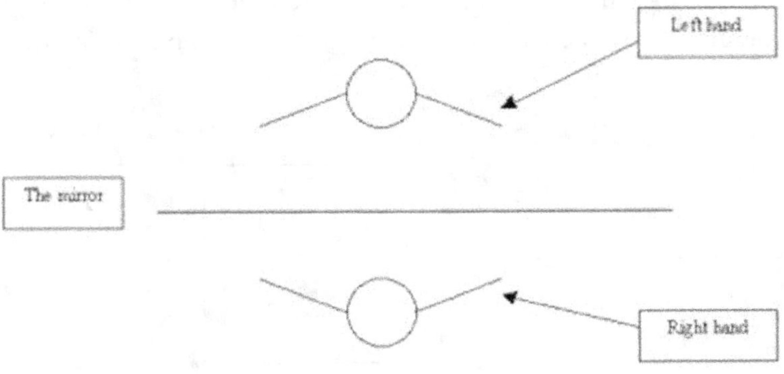

So the number one metaphysically acts as a mirror causing ∞/1 to become 1/∞:

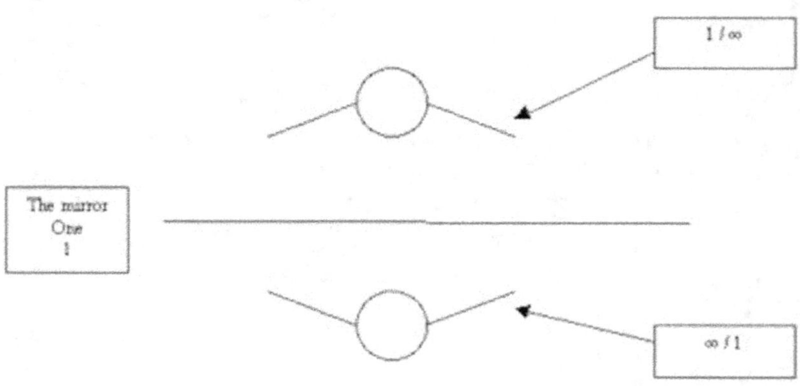

In the Aristotelian system, the whole is found outside the individual and is composed of multiplicity, multiplicity of individualities found 'outside' the individual. As such multiplicity abounds.

As such we obtain the infinite multiplicity of the whole as expressed by the divisibility of the infinite/multiplicity as seen through the eyes of the whole/singularity. Mathematically one might express the concept as:

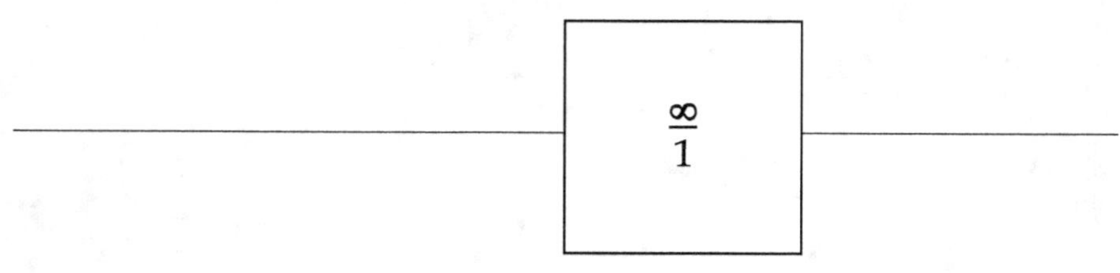

In the Kant/Hegelian system, the whole is found inside the individual and is likewise composed of multiplicity. Rather than multiplicity of individualities found 'outside' the individual, however, multiplicity is composed of the multiplicity of experience and thought found 'inside the individual. As such multiplicity abounds.

As such we obtain the infinite multiplicity of the whole as expressed by the divisibility of whole/singularity (1) as seen through the eyes of the infinite/multiplicity itself. Mathematically one might express the concept as:

$$\frac{1}{\infty}$$

Such concepts as:

$$\frac{\infty}{1} \qquad \frac{1}{\infty}$$

Need a little attention before we move into examining the concept of the Einsteinian 'i' and its relationship to metaphysics.

Two additional concepts emerge from the equation, v = d / t and the equation v' = t / d:

$v = d/t$	$v' = d/t$
$vt = d$	$v't = d$
$v = k$	$v = k$
$k = d/t$	$k = t/d$
$k = 1^*$	$k = 1^*$
$1 = d/t$	$1 = t/d$
If d and t are equivalent 'substances' as previously suggested, we then find:	If d and t are equivalent 'substances' as previously suggested, we then find:
If $d = 0$, then $t = 0$ and	If $d = 0$, then $t = 0$ and
$1 = 0/0$	$1 = 0/0$
If $d = \infty$, then $t = \infty$ and	If $d = \infty$, then $t = \infty$ and
$1 = \infty / \infty$	$1 = \infty / \infty$

We thus find we must not only examine the metaphysical significance of:

$$\frac{\infty}{1}$$

$$\frac{1}{\infty}$$

But we must also examine the metaphysical significance of:

$$1 = 0/0$$

$$1 = \infty/\infty$$

13. 0

$v = d/t$

$vt = d$

$v = k$

$kt = d$

$k = 1^*$

$t = d$

* Where k is understood to be the constant of physicality, the constant variable, the variable constant

$v' = t/d$

$v'd = t$

$v' = 1/k$

$1d/k = t$

$k = 1^*$

$d = t$

* Where k is understood to be the constant of physicality, the constant variable, the variable constant

Metaphysically speaking the concept implies equality of the abstractual values of time and distance/space as is the case of all abstractions found within the purity of abstraction.

$$t = \infty$$

$$d = \infty$$

$$v = \text{constant variable} = k = 1$$

$\frac{vt}{1} = \frac{d}{1}$	$\frac{1}{vt} = \frac{1}{d}$
$\frac{k\infty}{1} = \frac{\infty}{1}$	$\frac{1}{k\infty} = \frac{1}{\infty}$
$\frac{\infty}{1} = \frac{\infty}{1}$	$0 = 0$
What does it mean when we suggest dividing infinity by one, dividing infinity by the whole? It means accepting endless division of 'what is':	What does it mean when we suggest dividing one/the whole by infinity, dividing one by the infinite? It means infinite division of 'what is' creating infinite multiplicity of 'what is':
It means infinite creation of knowledge/newness is'. It implies a location for the future in the process of 'being created'. Growth through: What could be	It means retention of 'what is' as 'what It implies a location for the past in the process of being the past No growth to: What is What was What will be

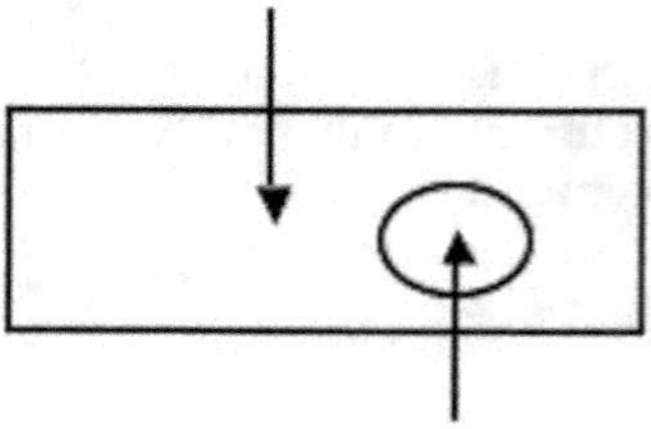

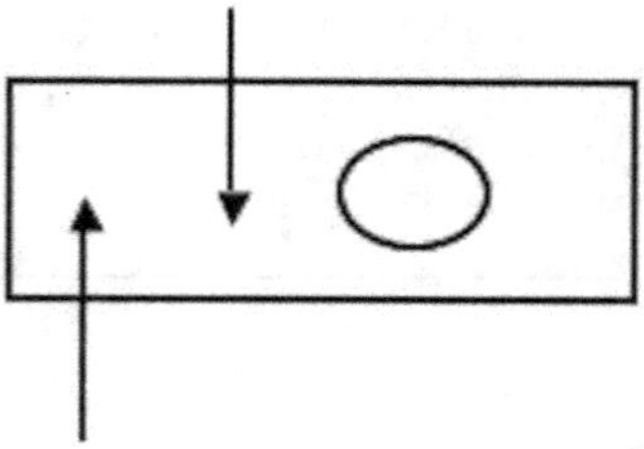

So it is ∞ / 1 becomes an issue while simultaneously becoming understandable as the representative of multiplicity.

<table>
<tr><td colspan="2">

$t = 0$

$d = 0$

$v = \text{constant variable} = k = 1$

</td></tr>
<tr><td>

$\frac{vt}{1} = \frac{d}{1}$

$\frac{k0}{1} = \frac{0}{1}$

$0 = 0$

What does it mean when we suggest dividing nothingness by one, dividing nothingness by the whole?

It means a potential for the 'could be', which as of yet is not nor is the 'could be' an absolute 'will be'.

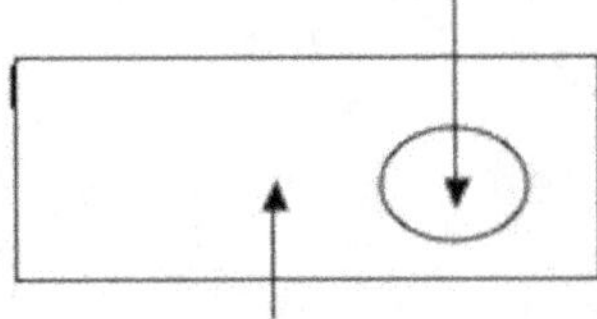

It means infinite creation of knowledge/newness.

It implies a location for the future in the process of 'being created'.

Growth: What could be

</td><td>

$\frac{1}{vt} = \frac{1}{d}$

$\frac{1}{k0} = \frac{1}{0}$

$\infty = \infty$

$\frac{\infty}{1} = \frac{\infty}{1}$

What does it mean when we suggest dividing one/the whole by one, dividing one by nothingness?

It means no division of 'what is' but rather it means the potential to make smaller the ratio of the singularity of multiplicity of what is to summation of the multiplicity of 'what is'

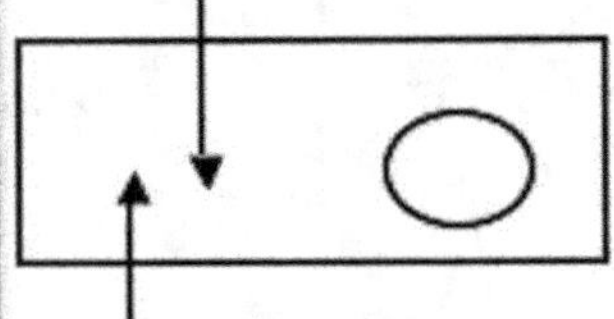

It means retention of what is as 'what is'.

It implies a location for the past in the process of being the past

**No growth: What is
What was
What will be**

</td></tr>
</table>

How is it one can metaphysically divide by zero when mathematically such an operation is defined out of existence?

It is not that dividing by zero is impossible but rather dividing by zero is defined by mathematicians as being an irrational act since in the minds of mathematicians, zero does not exist as a functional concept since nothingness is perceived as 'nothing' versus being perceived as 'something'.

It is the metaphysicians who must address such a complex and revolutionary concept as nothing being 'something' and thus having an active form of action versus our presently perceived perception of nothing being 'nothing' and thus having simply a passive form of existence.

This change of perception regarding an active functionality of zero sets the perception of zero on its head just as Copernicus and Kant set the perception of centricism and observation upon their heads.

Part of understanding the new metaphysical perception is the understanding regarding the concept of 'what could be' being a means of growth of 'what is'. As such the progression:

100 / ∞	=	0
...		
100 / ∞	=	1/ ∞
...		
...		
100 / 1000000	=	.0001
100 / 100000	=	.001
100 / 10000	=	.01
100 / 1000	=	.1
100 / 100	=	1
100 / 10	=	10
100 / 1	=	100
100 / .1	=	1000
100 / .001	=	10000
...		
...		
100 / 1/∞	=	∞
...		
100 / 0	=	∞

… becomes a representation of the growth of the whole of 'what is' becoming a new 'whole' more diverse in its multiplicity of 'what is' without the limitations of time and space interfering with the whole being just what it is, the 'whole', without the characteristics of sequential development placed upon it which space/distance and time place upon the sequential development of events found to emerge within the universal fabric created by space/distance and time found within the realm of the physical/the universe – process itself as so perceived from within the realm of the physical.

The concept of nothing-ness not only existing but existing as a fully functional 'something' is full addressed in Volume 12: The Error of Heidegger.

So it is we once again see ∞ / 1 and 1 / ∞ becoming issues while simultaneously becoming understandable as the representative of multiplicity.

What of the representation of singularity? The representation of singularity becomes the whole, the number one, 1.

If infinity, ∞/1 and 1/∞, are the representation of multiplicity and if one, 1, is the representation of singularity/the whole, then what is zero, 0?

For now we will address zero in the more elementary sense as being the beginning of it all as demonstrated by the two graphs.

Having given some consideration to both one and zero, let's examine the metaphysical implications of the mathematical representations of multiplicity: ∞ / 1 and 1 / ∞.

14. Introduction to ∞ / 1 and 1 / ∞

The fractions, ∞ / 1 and 1 / ∞, are the ratios of the total of the parts to the whole and the whole to the total of the parts:

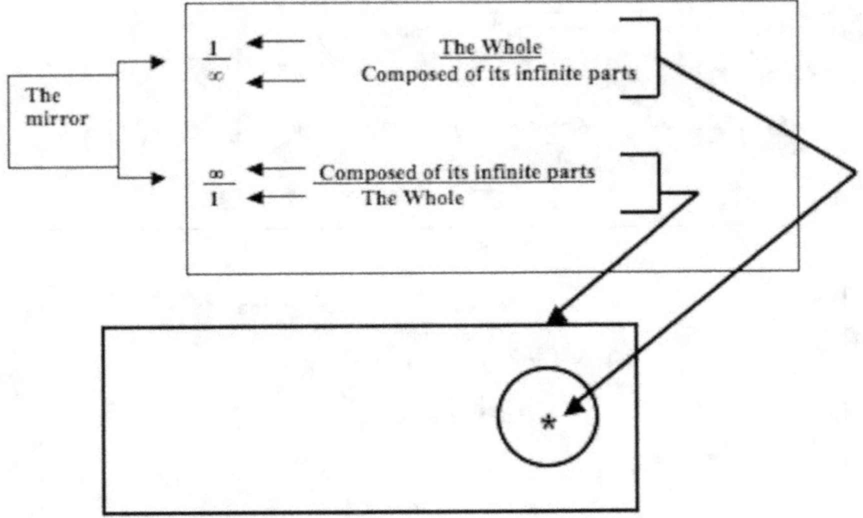

The mirror inverts the image. In this diagram the mirror converts ∞ / 1 to 1 / ∞:

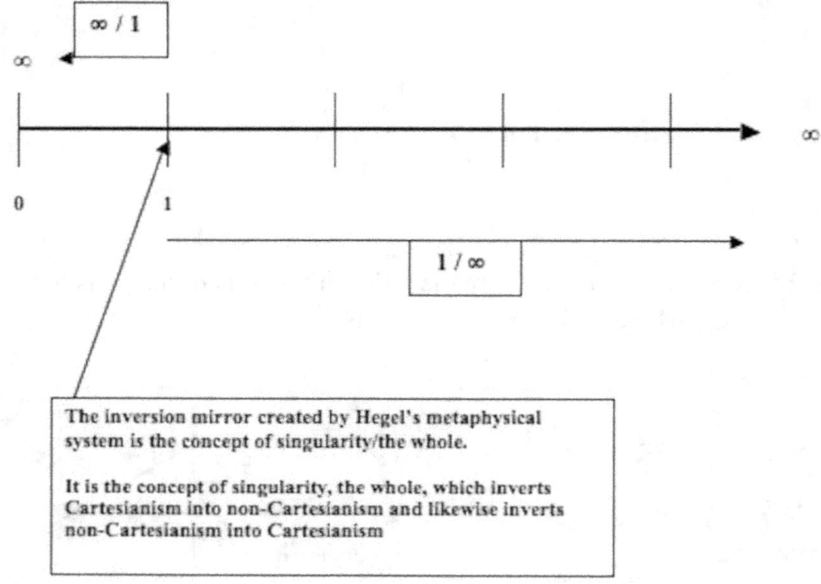
∞ / 1
∞
∞
0
1
1 / ∞
The inversion mirror created by Hegel's metaphysical system is the concept of singularity/the whole.
It is the concept of singularity, the whole, which inverts Cartesianism into non-Cartesianism and likewise inverts non-Cartesianism into Cartesianism

For the physical:

The mirror creates the image inversion
As such
'left = right' becomes 'right = left'

•

For the abstract

The mirror creates the image inversion

'time depends upon distance' becomes 'distance depends upon time'

•

For the whole

'the abstract depends upon the physical' becomes 'the physical depends upon the abstract'

through

"the real' = 'the real illusion" becomes "the real illusion' = the real"

15. ∞ / 1

$t = \infty / 1$

$d = \infty / 1$

v = constant variable = $k = 1$

$\frac{vt}{1} = \frac{d}{1}$	$\frac{1}{vt} = \frac{1}{d}$
$\frac{k\infty}{1} = \frac{\infty}{1}$	$\frac{1}{k\infty} = \frac{1}{\infty}$
$\frac{1\infty}{1} = \frac{\infty}{1}$	$1 + \infty = 1 + \infty$
$\frac{\infty}{1} = \frac{\infty}{1}$	and / or $\frac{1}{\infty} = \frac{1}{\infty}$

What does it mean when we suggest dividing one/the whole/singularity by infinity/multiplicity?	**What does it mean when we suggest dividing infinity/multiplicity by one/the whole/singularity?**
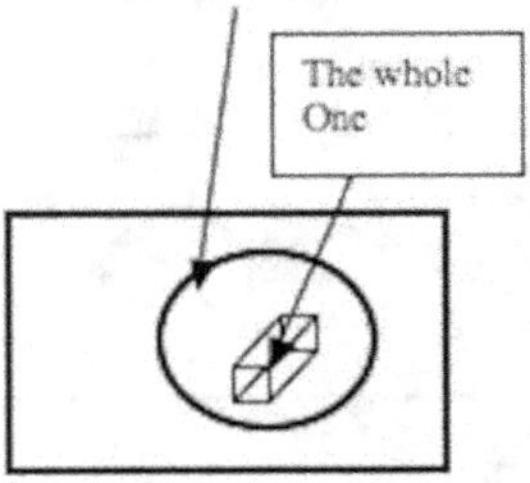 The whole One	The whole One
It means:	**It means:**
Endless division of 'what is':	**Endless division of what is creating the infinite multiplicity of 'what is':**
Retention of 'what is' as 'what is'.	**Infinite creation of knowledge/newness.**
It implies a location for the past in the process of being the past	**It implies a location for the future in the process of 'being created'.**
No perceived growth to: What is / What was / What will be	**Growth through:** What could be

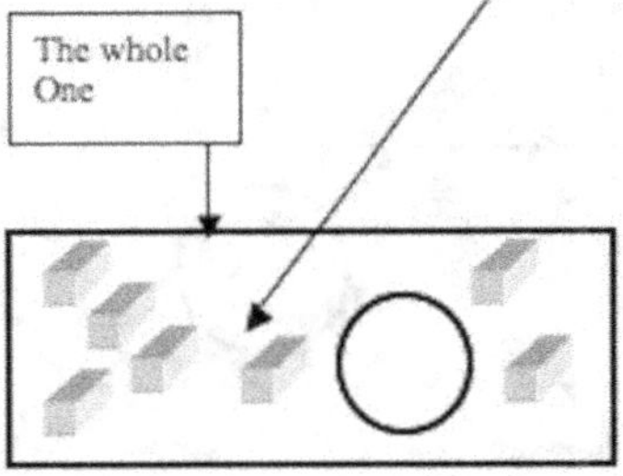

So it is ∞ / 1 becomes an issue while simultaneously becoming understandable as the representative of multiplicity.**1 / ∞**

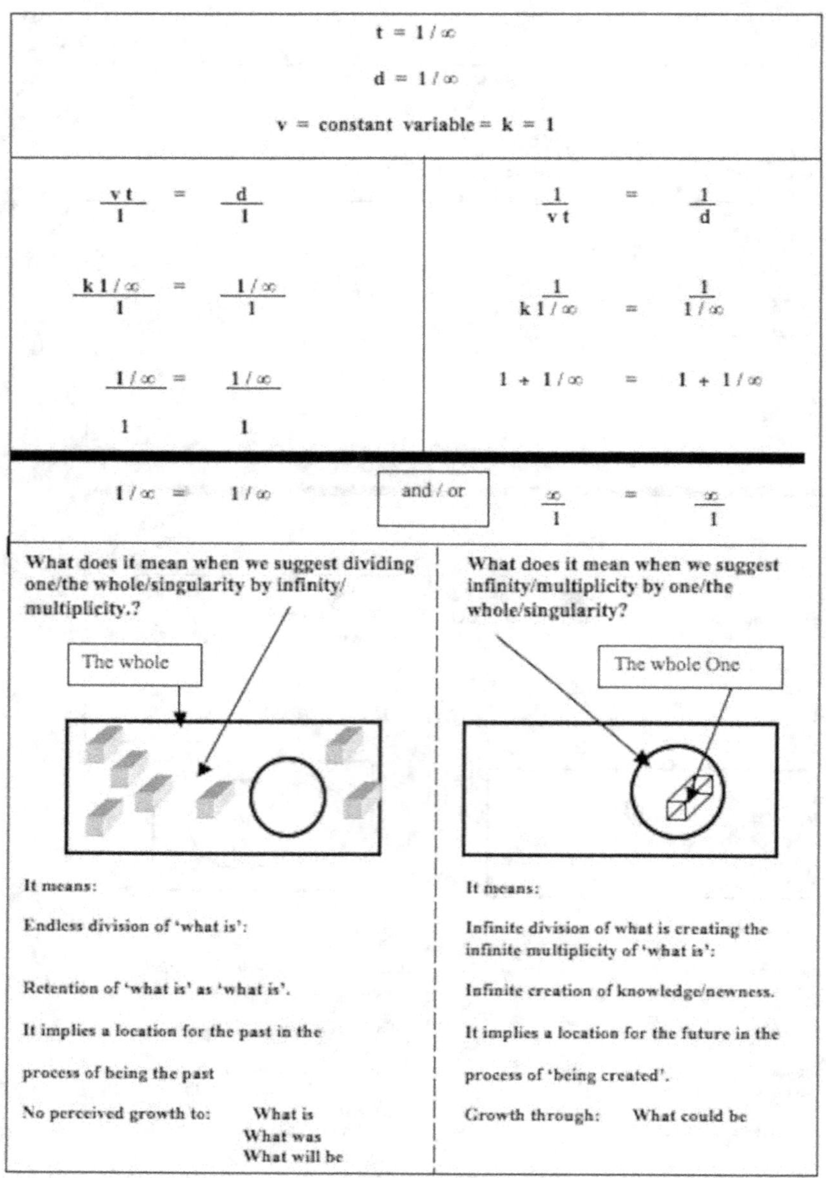

So it is 1 / ∞ becomes an issue while simultaneously becoming understandable as the representative of multiplicity.

So it is the 1st mirror becomes significant. One divided by infinity and infinity divided by one become the issue not as separate concepts but rather as identical concepts from which emerges growth of perceived ‘no growth’ found immersed ‘within’ the void of time through ‘what could be’ found immersed ‘within’ time becoming ‘what is’.

16. 1 = 0 / 0

$$t = 0$$

$$d = 0$$

v = constant variable = k = 1
v' = constant variable = 1 / k = 1

$v = \frac{d}{t}$	$\frac{1}{v} = \frac{t}{d}$
$k = \frac{0}{0}$	$\frac{1}{k} = \frac{0}{0}$
$1 = \frac{0}{0}$	$1 = \frac{0}{0}$

What does it mean when we suggest dividing nothingness by itself, nothingness?

It means endless division of 'what is':

Infinite creation of knowledge/newness.

It implies a location for the future in the process of 'being created'.

Growth: **What could be**

It means no division of 'what is':

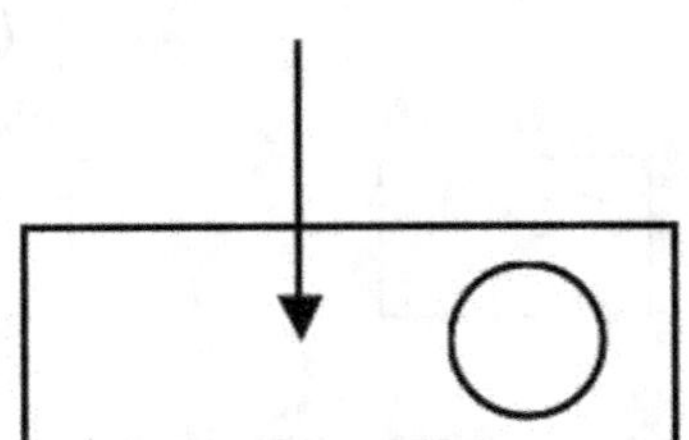

Retention of 'what is' as 'what is'.

It implies a location for the past in the process of being the past

No growth: **What is**
What was
What will be

Nietzsche: God is dead.

Which came first - the chicken or the egg?

Which came first – God or creation?

The old paradox stems from a linear versus a quadratic form of thinking.

A linear form of thinking emerges from a mathematical form of logic where the equation is formed from perceptions to the 1st power which leads from here to there rather than a form of thinking where the equation is formed from perceptions to the 2nd and 3rd powers.

Linear progressions have a beginning even if the beginning lies infinitely far 'behind'.

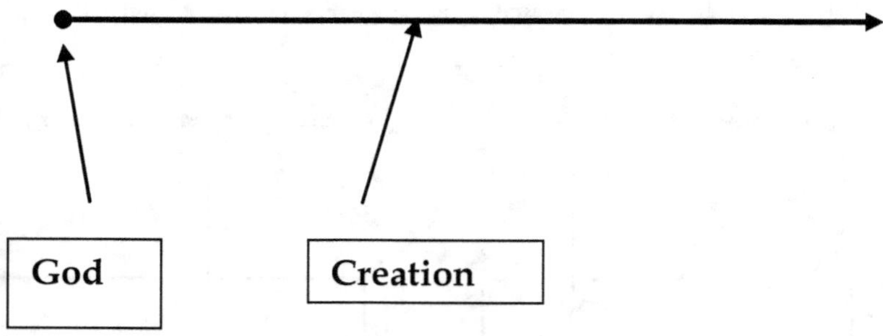

Quadratic progressions have no beginning and thus no 'solution' to the chicken/egg dilemma.

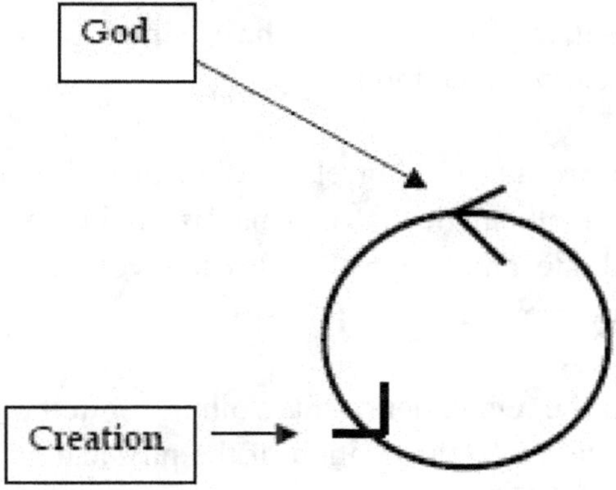

Why are the arrows not placed symmetrically?

The arrows are placed off symmetrical balance to indicate the 'beginning' 'end' may not be the big picture rather 'beginning' 'end' is simply our perception of the big picture.

This then brings us to the interrelationship of the concrete and the abstract as symbolized mathematically:

$E = mc2$	$E - mc2 = 0$
$E = mc2$	$E - mc2 = 0$
Energy Matter	The Concrete The Abstract
Energy and matter become the constants	**The concrete and the abstract become the constants**
Conservation of matter and energy:	**Conservation of the physical and the abstract:**
Matter and Energy cannot be created or destroyed.	**The physical and the abstract cannot be created or destroyed**

The interrelationship between matter and energy is clearly indicated mathematically and is clearly understood.

The interrelationship of the physical and the abstract is likewise clearly indicated but not so clearly understood.

The understanding regarding the interrelationship of the concrete and the abstract as indicated mathematically is what is to be explored in the remainder of this volume and the two following volumes: Volume 13: Russell and Volume 14: Heidegger.

As we shall see within the remainder of this volume, both the Conservation of matter and energy and the Conservation of the physical and the abstract are equally erroneous in their perceptual interpretation.

But what of the concepts of mirror images, constant variability, variable constancy, and constant consistency of change?

Let's begin be stating that the concrete, as represented by the equal sign in the equation E = mc2, has a mirror image of itself.

Likewise we can state that the abstract, as indicated by the equal sign in the equation E – mc2 = 0, has a mirror image of itself. In essence we have two mirrors functioning in conjunction one with the other.

It will help us to understand such statements if we understand that before the Kant/Hegel/Nietzsche perceptions immerged, the perception of reality was:

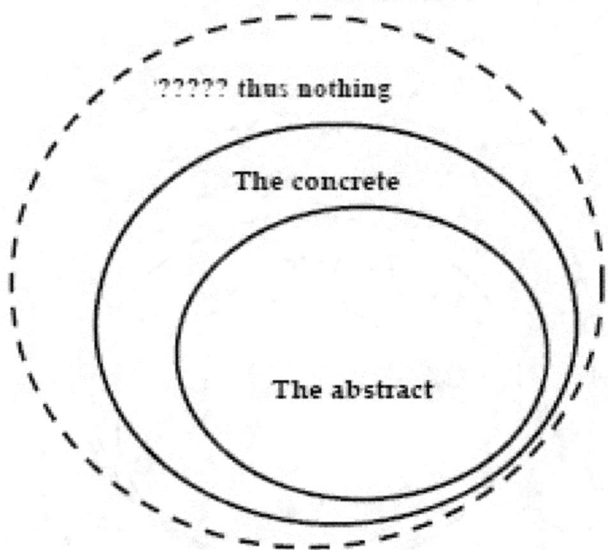

The abstract was a part of the physical. The abstract was to be found 'within' time and space which in turn was within the concrete/physical.

Following the development of the Kant/Hegel/Nietzsche perceptions, the competing perception of reality became:

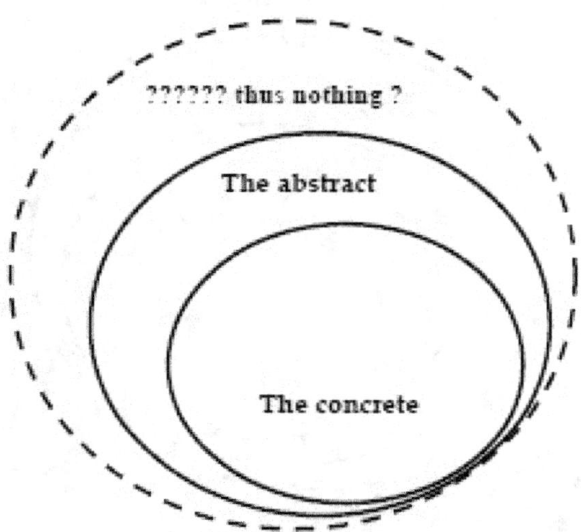

The physical was a part of the abstract. The physical was to be found 'within' the abstract and thus time and space being a part of the physical were aspects of the abstractual.

And so it was Nietzsche could declare the death of God for 'we' no longer 'needed' God to create change since change was an element of ourselves.

Regardless of the perception of change, the overall picture in fact did not change: The physical remained imbedded in non-relativism in terms of the significance of one entity over another and the metaphysical remained imbedded in relativism in terms of significance of one over another.

With Einstein came the inversion of such perceptions.

Einstein did to the significance of the interrelationship of physical and the abstract what Copernicus did to the interrelationship of the center and man and what Kant did to the interrelationship of the physical and observation.

As such the physical metaphysically became relative in terms of the significance of one entity to another as opposed to being imbedded in non-relativism in terms of the significance of one entity over another and the abstractual metaphysically became non-relativistic in terms of the significance of one entity to another as opposed to being relative in terms the significance of one entity to another.

In short: With the advent of Einstein's theory of relativity, relatively speaking we are all the same in terms of the 'greater' abstractual significance and we are not the same in terms of the 'lesser' physical significance.

Some entities represent the pillars and some entities represent the ones who sum up what the pillars are holding up. Whether one is the pillar or the summation of what it is the pillar represents, all are equally significant since the summation of the pillar cannot occur without the pillar.

Einsteinian mathematics of relativism forged a new perception, which metaphysically, from an Aristotelian metaphysical point of view, now becomes:

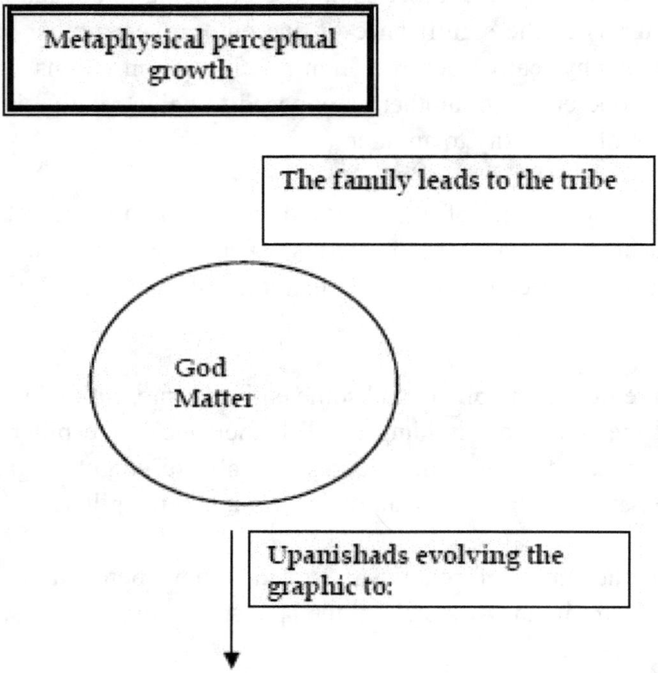

Metaphysical perceptual growth

The family leads to the tribe

God
Matter

Upanishads evolving the graphic to:

The tribe leads to the nation leads to the planet

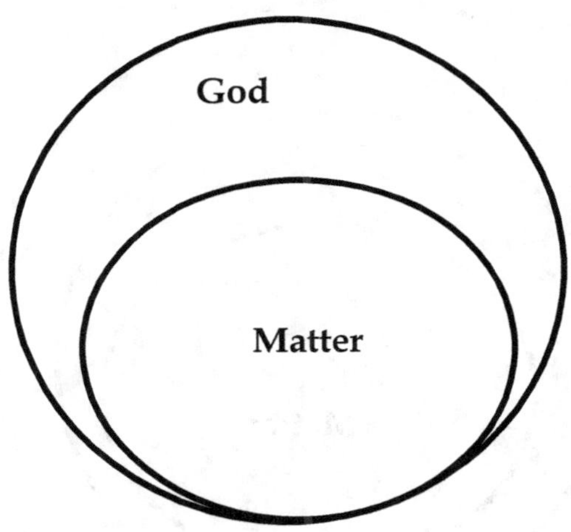

Kant/Hegel/Nietzsche evolving the graphic to:

The planet leads to the galaxy leads to the universe

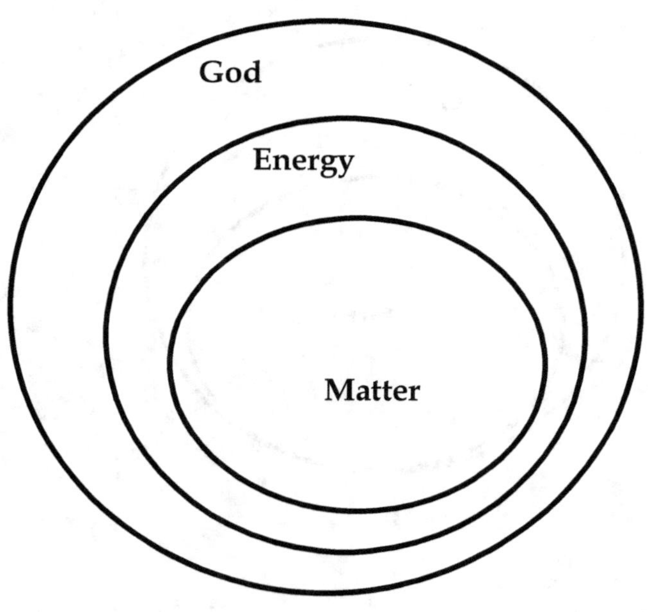

Einstein evolving the graphic to:

The universe leads to other universes

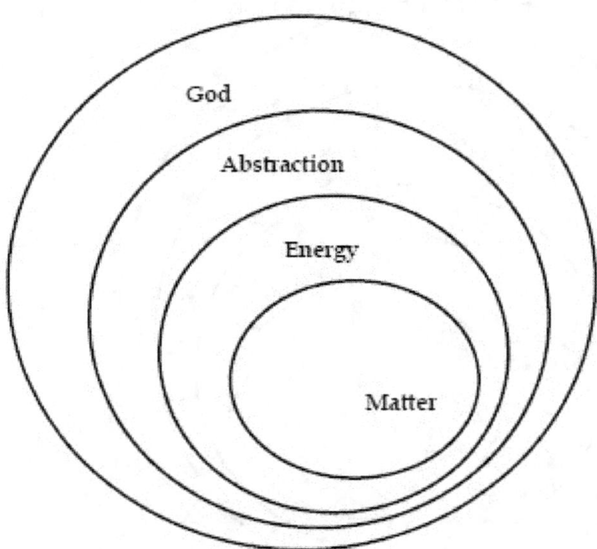

Metaphysics evolving the graphic to:

The other universes lead to what lies 'between' the universes

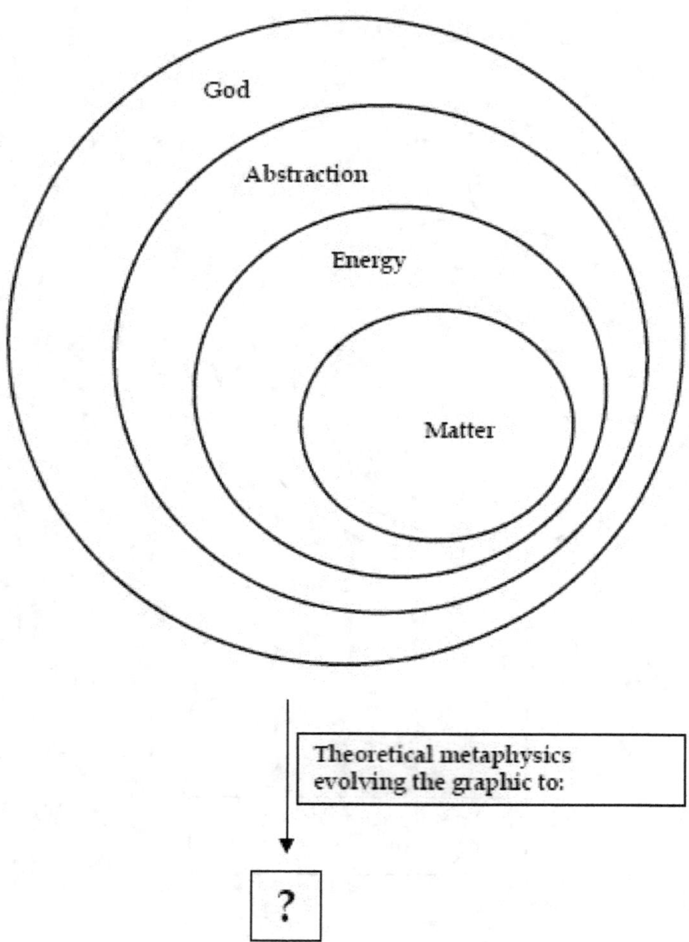

Reinterpreting the preceding series of graphics from a present day cosmological point of view we obtain:

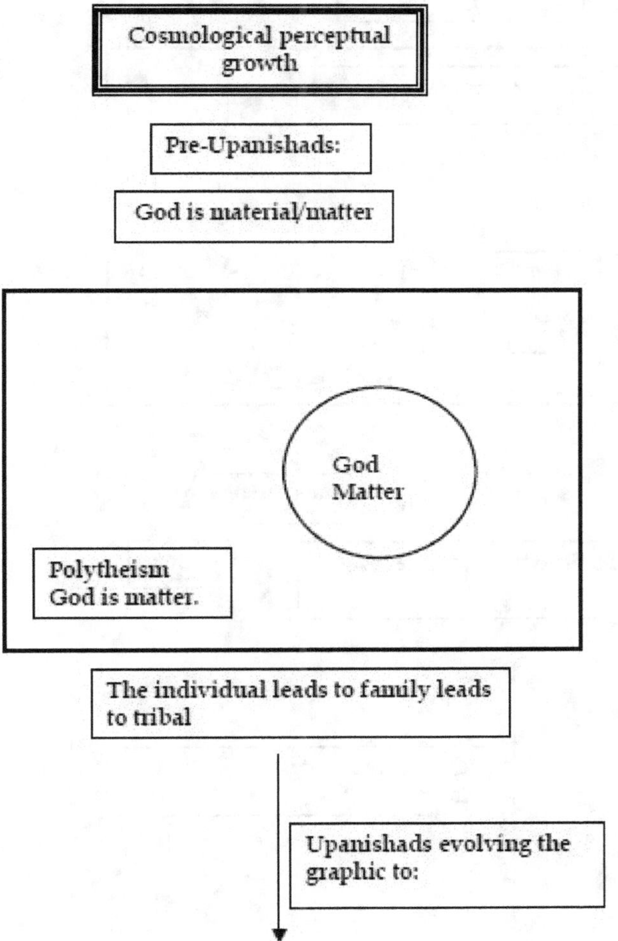

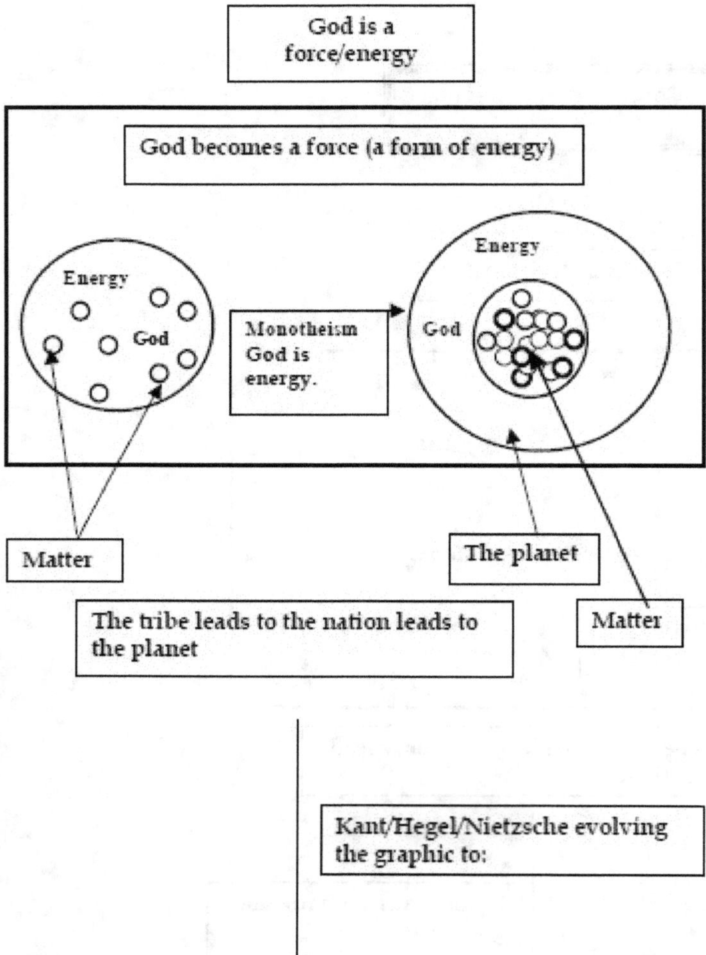
God is a force/energy
God becomes a force (a form of energy)
Energy
God
Monotheism God is energy.
Energy
God
Matter
The planet
The tribe leads to the nation leads to the planet
Matter
Kant/Hegel/Nietzsche evolving the graphic to:

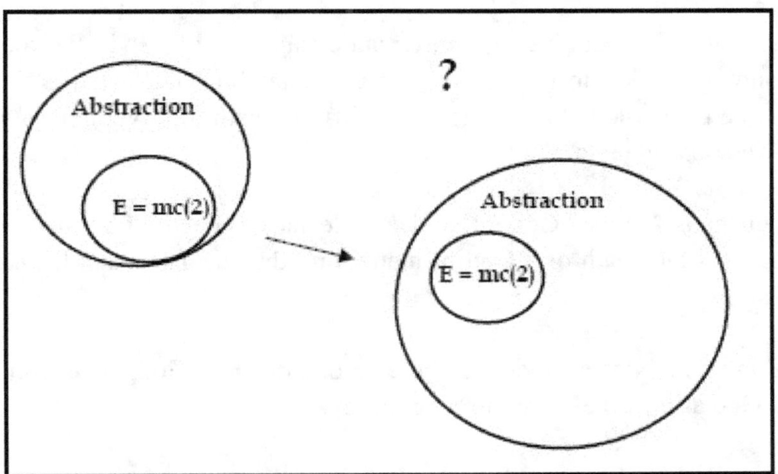

Leaving us with the question: What is '?'.

The flow mathematically is depicted as:

E = mc(2)

E – mc(2) = 0

(E – mc(2)) – 0 = **?**

There are four metaphysical elements to address within this flow. The four metaphysical elements are:

1. The flow itself
2. The understanding of what it means to subtract mc(2) from E

3. The understanding of what it means to subtract ‘0’ from E – mc(2)
4. The understanding of ‘?’

It is the understanding of these four metaphysical elements which we are to address within the remainder of this volume and throughout volumes 13 - 22.

We are going to examine the next understanding of God, the next understanding of God to come, the next understanding of what it is God may be. We have a long way to go and a difficult path to travel before we come to less taxing texts.

The Upanishads declared God’s first death, declared a death of a ‘physical’ God, declared the death of a God of matter and thus declared the death of polytheism.

Hegel along with Nietzsche declared the second death of God, declared the death of God as a ‘force’ composed of energy.

Einstein inadvertently showed us the way to understand the new form of god as a God of abstraction.

The point is: we can now understand we have just begun our understanding of the big picture as depicted in Aristotelian format:

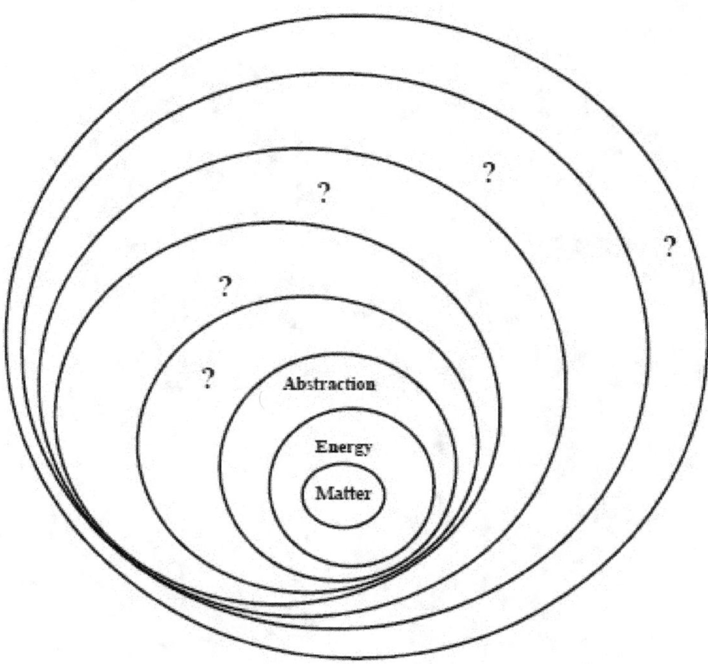

We have begun to examine the characteristics of the first three inner shells.

We have much to learn regarding the inter and intra relationship existing between the three inner shells: matter, energy, and abstraction before we can begin an in-depth examination regarding the possibility of a fourth shell let alone begin examining the fifth, sixth, seventh, etc shells.

Do not confuse the concept of shells four, five, six, etc with the concepts of multiple dimensions to our universe. Multiple dimensions to the universe and multiple shells to reality apply to different aspects of the whole.

We now come to the fourth concept regarding Newtonian physics and metaphysics:

17. $1 = \infty / \infty$

$$t = \infty$$

$$d = \infty$$

v = constant variable = k = 1
v' = constant variable = 1 / k = 1

$v = \frac{d}{t}$	$\frac{1}{v} = \frac{t}{d}$
$k = \frac{\infty}{\infty}$	$\frac{1}{k} = \frac{\infty}{\infty}$
$1 = \frac{\infty}{\infty}$	$1 = \frac{\infty}{\infty}$

What does it mean when we suggest dividing infinity by itself?

It means endless division of 'what could be':

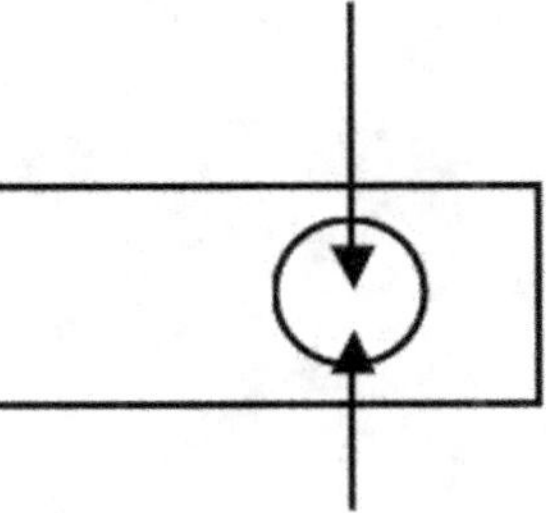

It means:

Infinite creation of knowledge/newness.

It implies a location for the future in the process of 'being created'.

Growth: What could be

It means no division of 'what could be':

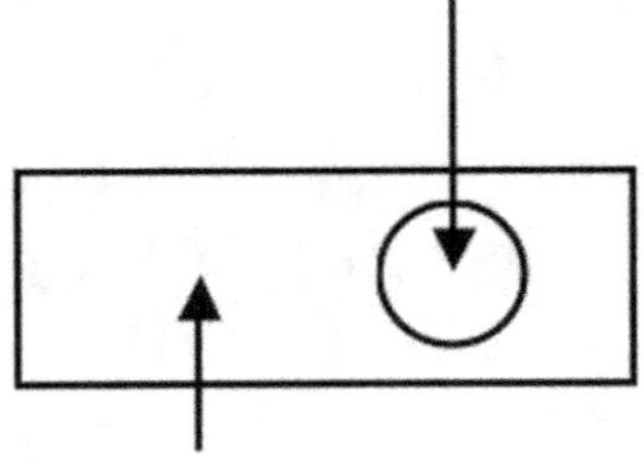

It means:

Retention of 'what is' as 'what is'.

It implies a location for the past in the process of being the past

**No perceived growth to: What is
What was
What will be**

Before we can get to the significance of the speed of light in terms of the equation, E = mc2, as it relates to metaphysics, we must examine the linear concept of velocity and the significance of velocity to matter and energy in terms of metaphysics. To do so requires an understanding regarding the metaphysical significance of 0 / ∞ versus ∞ / 0, 0 = 0, t / 1, and d / 1.

18. 0 / ∞ versus ∞ / 0

Zero divided by anything is itself/zero/potentiality. All 'things' have the potentiality of expanding into being 'greater' than what they are.

Anything divided by zero/potentiality is itself since potentiality is not but rather potentiality is the potential to be, is 'what could be'.

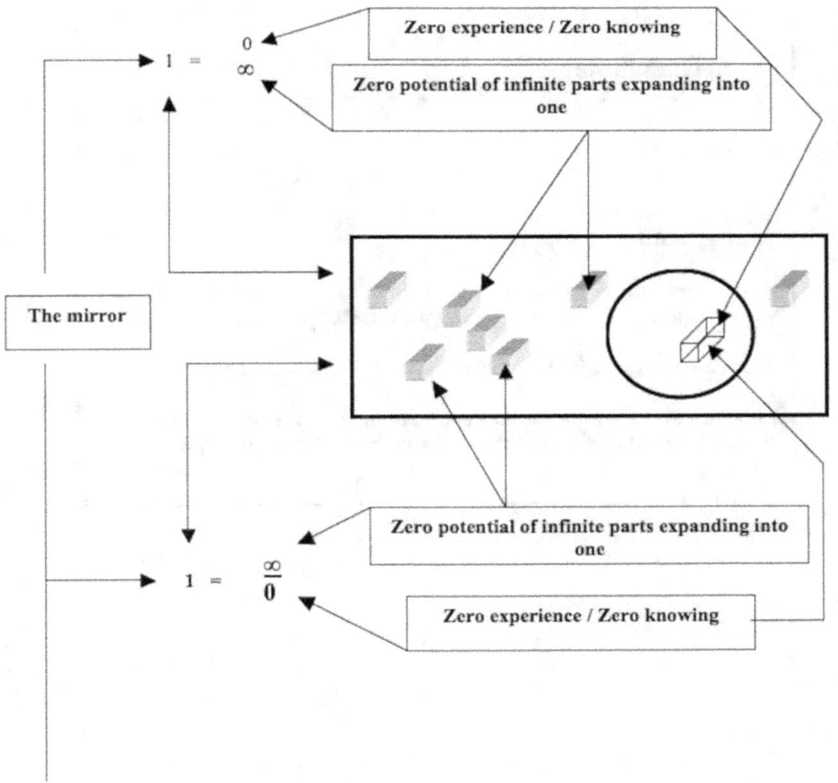

(Graphic continued on next page)

(Graphic continued from previous page)

By mathematical definition one cannot divide by zero. Metaphysically it is also irrational to divide by zero, however, metaphysically one can begin to understand the rationality of the statement that although one cannot divide by potentiality, dividing by zero represents a region of the whole of reality being able to be divided by itself for the whole of all realit yincludes potentiallity:

$$\text{The whole} = \sum \frac{\sum_{1}^{\infty}}{\sum_{1/\infty}^{1}} + 1$$

The implication:

> **The wole is greater than the sum of its parts for the whole is equal to the sum of its parts plus itself/the whole.**

The whole is an entity in itself as a summation of all its parts.

Likewise one, a whole part is a summation of its parts unable to be its complete self if any part of itself, if any experience of itself, is missing.

19. 0 = 0

$$v = d/t$$

$$d/t = d/t$$

$$(t)\,d/t = (t)\,d/t$$

$$d = d$$

The question becomes: What does d = d imply metaphysically? From our point of perception, d = d is taken as a truism that d = d but it can also be perceived that such a statement would be:

$$d_f = d_a$$

Where d sub 'f' equals functional distance located within functional time and d sub 'a' equals abstractual distance located within abstractual time.

Whether the relationship existing between functional time and abstractual time exists in one dimension located 'within' another or exists in one location 'outside' another makes no difference.

The relationship existing between the two forms of distance could graphically be demonstrated as:

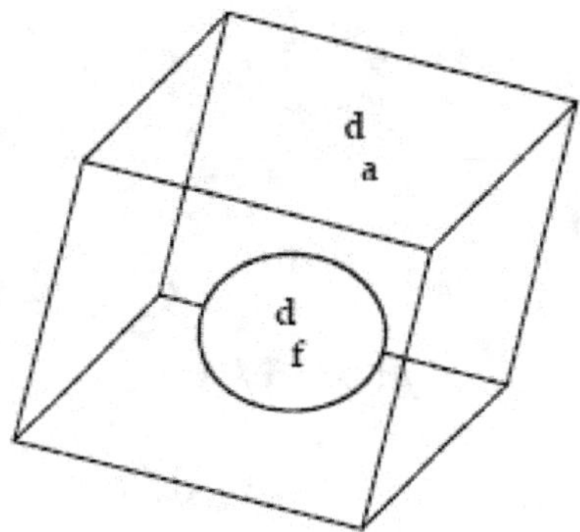

'd' sub 'f' is portrayed as existing in three dimensions, however, d sub 'f' could just as well be a function of four dimensions:

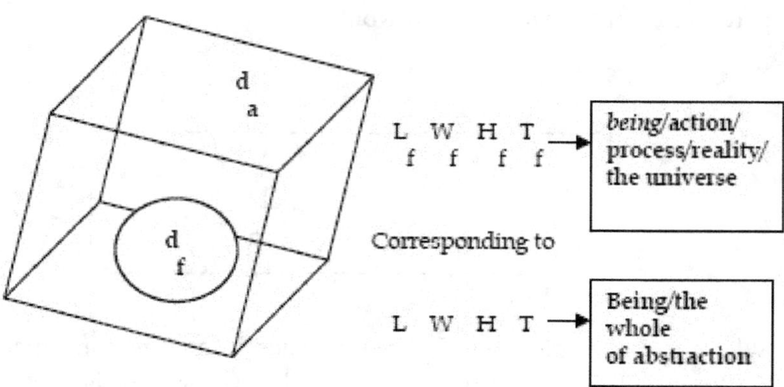

Zeno purposely deals with the positive real and inadvertently deals with the negative real. Thus Zeno deals with +/- L, +/- W, +/- H, +/- T, +/- D.

Einstein inadvertently deals with the imaginaries, deals with the 'i's', deals with the +/- iL, +/- iW, +/- iH, +/- iT, +/- D.

The understanding regarding the existence of positive and negative reals and the positive and negative imaginaries and the understanding of the inter-relationship of such existences is what metaphysics is in essence all about.

But metaphysics is more than just an examination and understanding of mathematical complexities; metaphysics is the understanding of the very 'whole' of it all, the understanding of +/- 'B' and +/- 'iB'

In actuality we have two graphics existing simultaneously, the real and the 'real illusion' which oscillate between being the real and the 'real illusion' to being the 'real illusion' and the real:

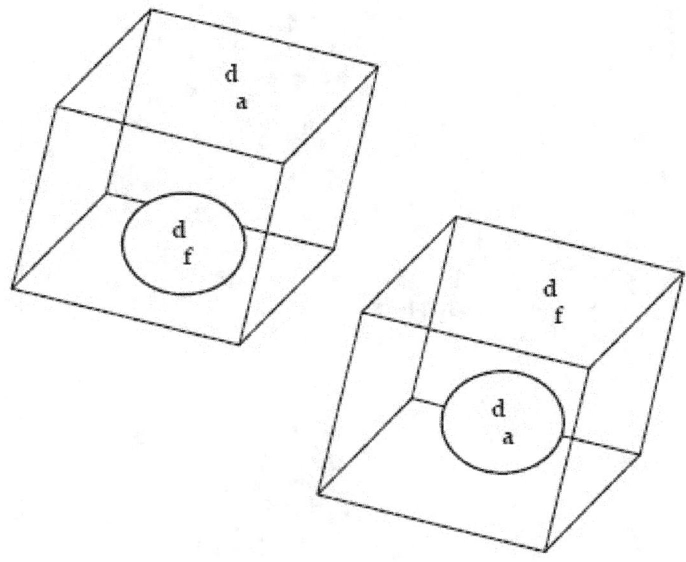

Within the graphic, d sub f represents functional distance found within the physical and d sub a represents non-functional distance found within the abstract. As such we obtain:

$$\frac{d_f}{t} = v$$

$$d_f = v\,t$$

$$d_f = \frac{d}{t}\,t$$

or

$$d_f = d_a$$

and

$$d_f = \frac{d_{a}\,t}{t}$$

We observe we have the same distance but now:

$$d_f = \frac{d}{t}\, t$$

Represents functional distance wrapped in time and:

$$\mathbf{1d_{f_1} = 1d}$$

'c' the coefficient of d equals the constant of singularity/'a' unit/one

In other words: One functional unit of distance directly corresponds to one unit of distance, which leads to an infinite number of universes as described by:

$$\mathbf{d_{f_\infty} = \infty\, d}$$

'c' the coefficient of d equals the constant of singularity/'a' unit/one

Even Epictetus acknowledged a location where distance existed as a non-functional abstraction:

'...you are infinitely small in terms of space but equal to God in terms of reason.'

The point is: Individuality exists to experience the individuality of each function of distance and each function of time.

Individuality builds a single/one location of infinite distant and time implies the ability to travel in an infinite number of directions in terms of time and distance.

The result is the development of one unique/individualistic entity of distance and time.

The representation of the single unique entity of infinite time is represented by t/1, infinite segments of time within the whole.

The representation of the single unique entity of infinite distance is represented by d/1, infinite segments of distance within the whole.

What is this concept of d/1 and t/1? What is this concept of infinite segments of distance and infinite segments of time existing as a whole?

20. Time and distance both divided by 1

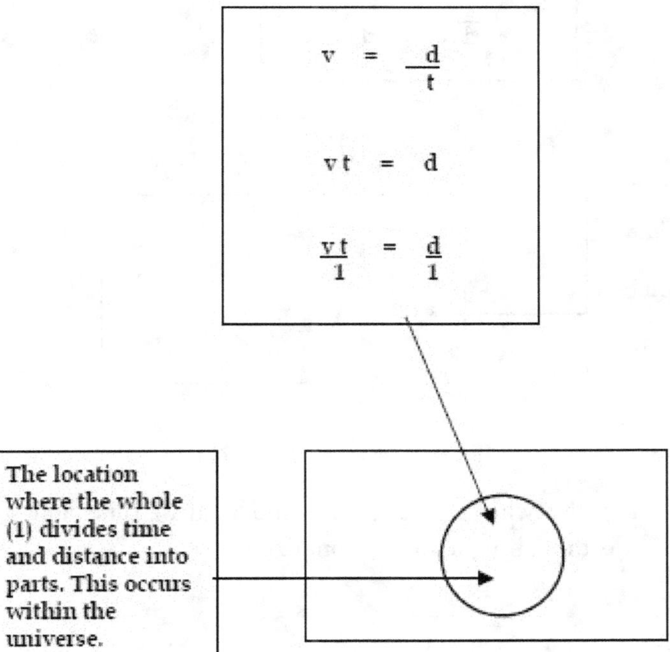
$v = \frac{d}{t}$
$vt = d$
$\frac{vt}{1} = \frac{d}{1}$
The location where the whole (1) divides time and distance into parts. This occurs within the universe.

v = a constant. 'v' is a variable in terms of numbers but 'v' is a constant in terms of ideas. 'v' is a constant in terms of abstractual functionality.

It is in the universe where time and distance are the fabric within which the individual, an active action of the present operating upon the future, is sectionalized.

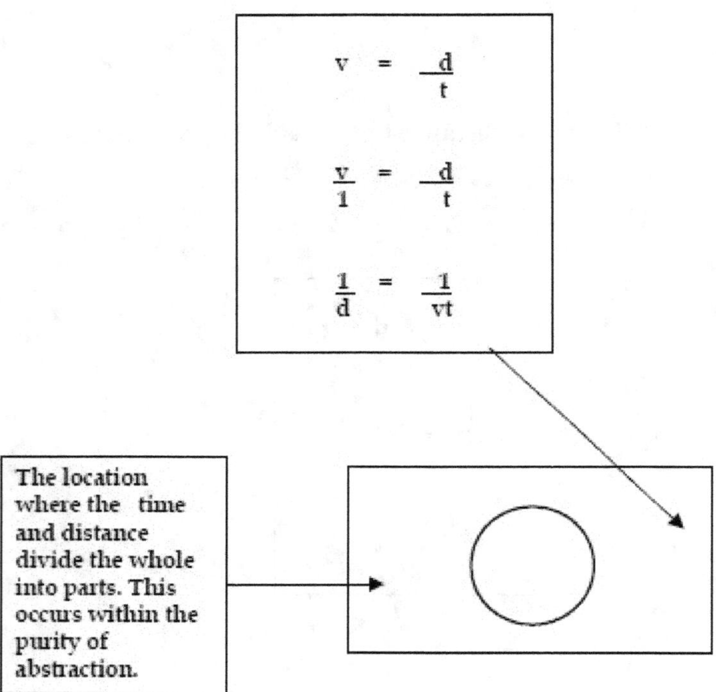

v = d/t
v/1 = d/t
1/d = 1/vt
The location where the time and distance divide the whole into parts. This occurs within the purity of abstraction.

It is the whole (1), the whole abstraction (the void of time and space) where abstraction is the fabric being sectionalized.

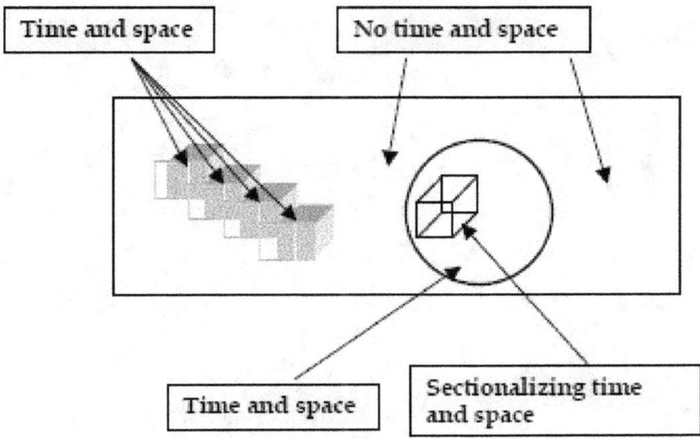

The result is a need for two separate independent yet simultaneously interdependent locations to explain the mathematics of multiplicity and seamlessness.

21. Knowledge: The universal building block

The universal building block, the primal atom of existence acting as the foundation of the new metaphysical system of the individual acting within God, panentheism, becomes intelligible through the understanding that the individual represents multiplicity of knowing, *being* represents action, process/reality/the universe, and God represents the singularity of knowing/ the whole and where the individual, *being*, and God are forms of 'knowing', *knowing*, and 'Knowing'.

The following progression helps explain knowledge being the universal building block:

The graphic:

?

Primal Atom

The universe

Earth

Atom

Subatomic particle

?

Becomes:

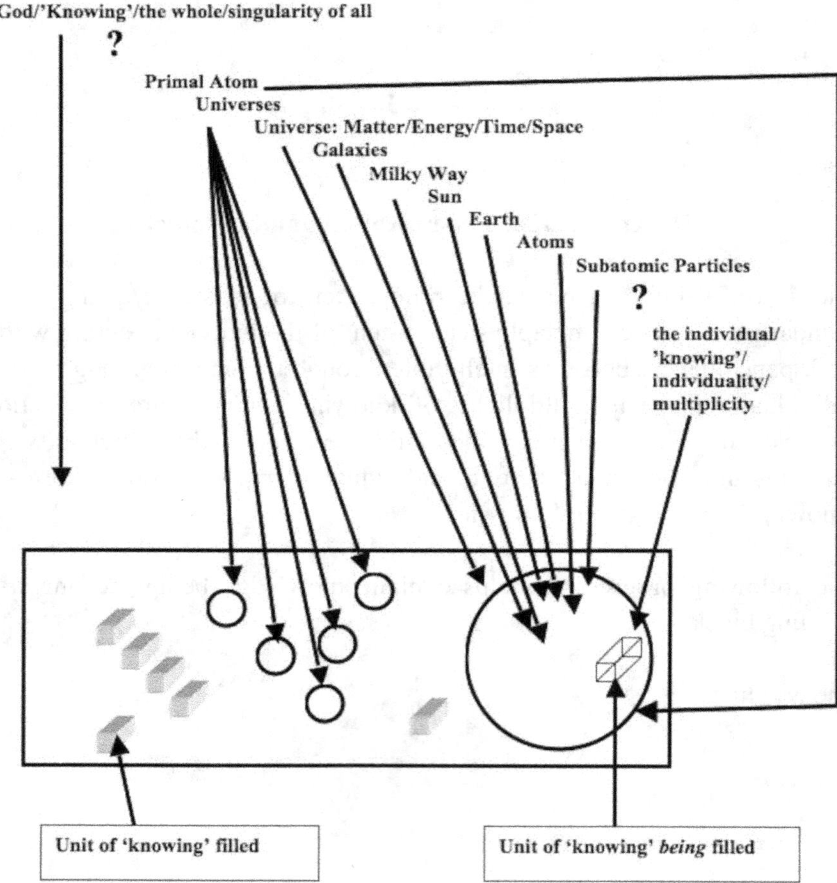

The universal building block becomes 'knowing'

The complete whole becomes 'Knowing'

In actually the progression becomes a quadratic relationship as opposed to a linear relationship and thus we obtain the graphic:

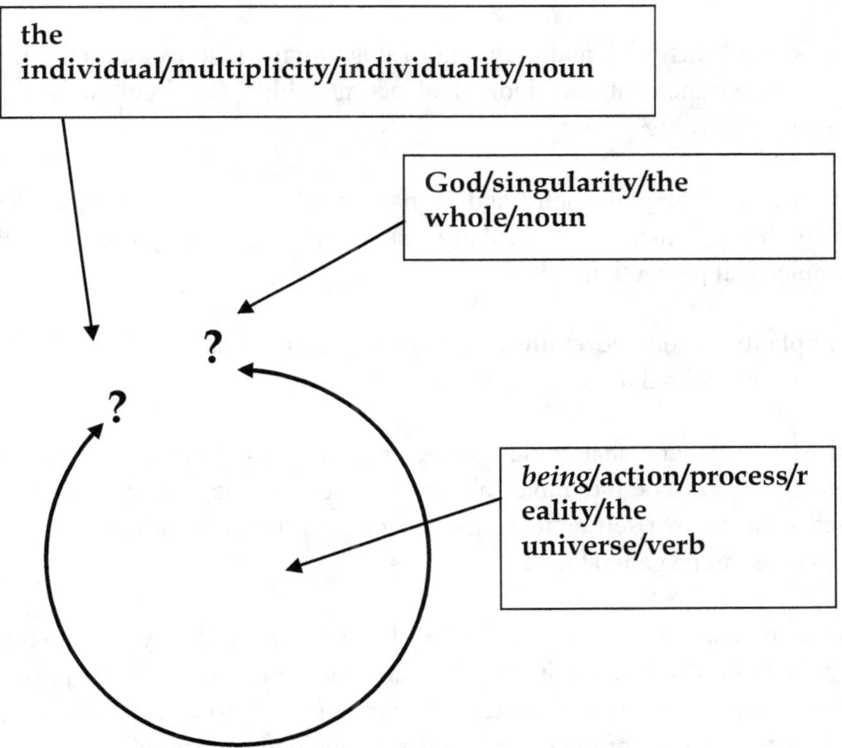

Therefore the two, ? and ? take on the appearance of being one in the same. The full understanding of the relationship therefore becomes twofold in nature:

the individual acting within God and the individual acting within God

or

'knowing' *knowing* 'Knowing' and 'Knowing' *knowing* 'knowing'

It is not the point of this work nor is it the point of this volume to examine the second half of each statement: the individual acting within God and 'Knowing' *knowing* 'knowing'. Regarding the second aspect of the relationship, the fields of Ontology and religion have said plenty.

The point of this work and the point of this volume is to examine the first half of the statement: the individual acting within God and 'knowing' *knowing* 'Knowing'.

It is the field of philosophy and in particular the field of Metaphysics, which have, practically speaking, said nothing in regards to this metaphysical perception.

Multiplicity of unique entities of knowing, multiplicity of individuality becomes the primal atom of the Whole of Knowing.

But what is the universal building block of universes? What is it that could be so small as to be the universal building block of the universe yet be so filled with its own self as to be the means by which the 'primal atom' of the whole can be generated?

As we shall see in Volume 12: The Error of Heidegger, it may very well be 'nothingness' itself acting in an active manner versus acting in the passive manner we have always perceived 'nothingness' to act which could reasonable account for the growth of the whole independent of time.

Perhaps the circle can help us see a useful progression, which might lead us to better understanding the universal building block of the whole of reality:

From a distance we have:

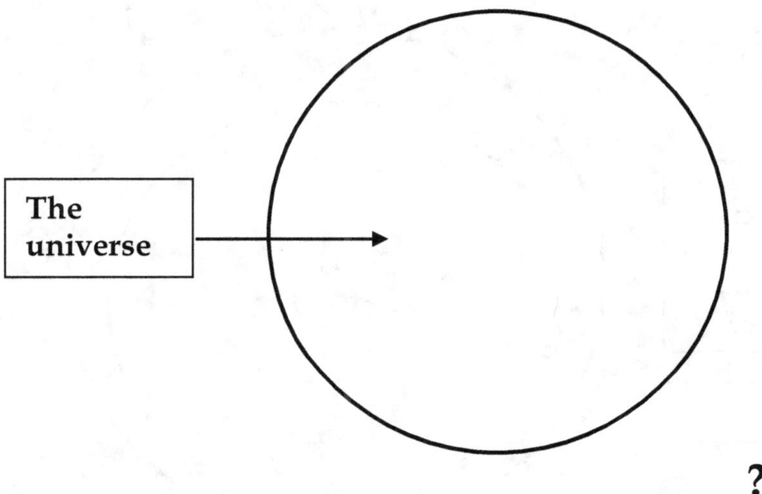

Zooming in we observe a progression of Aristotelian shells:

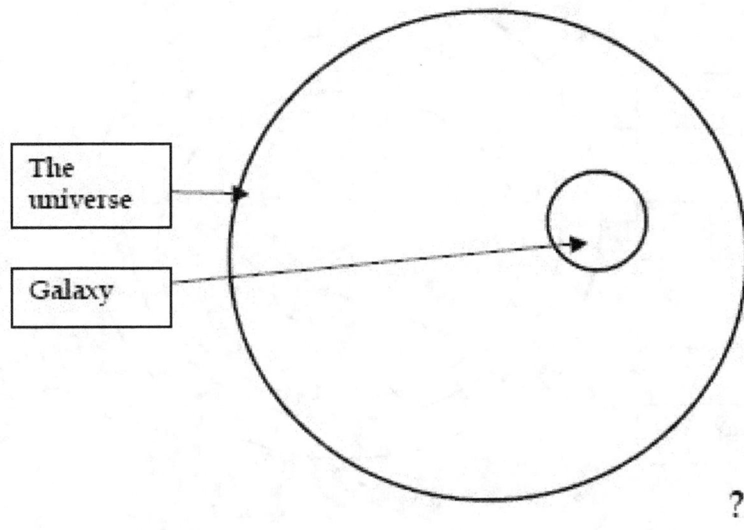

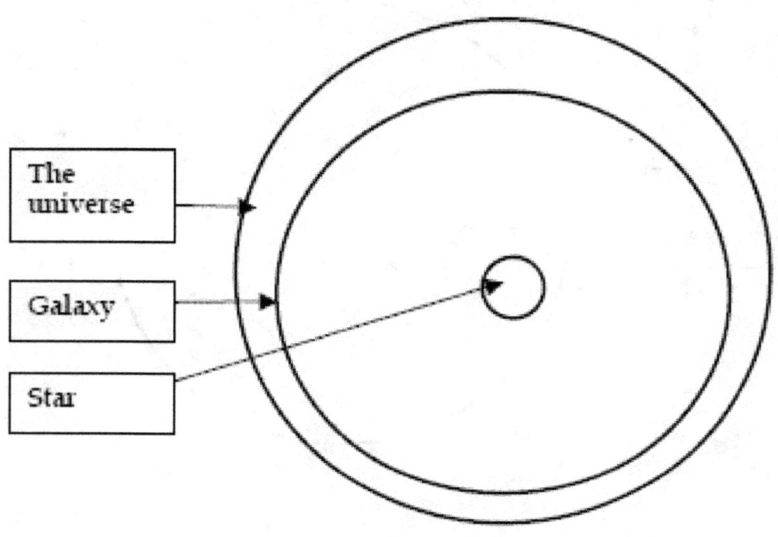
The universe
Galaxy
Star

?

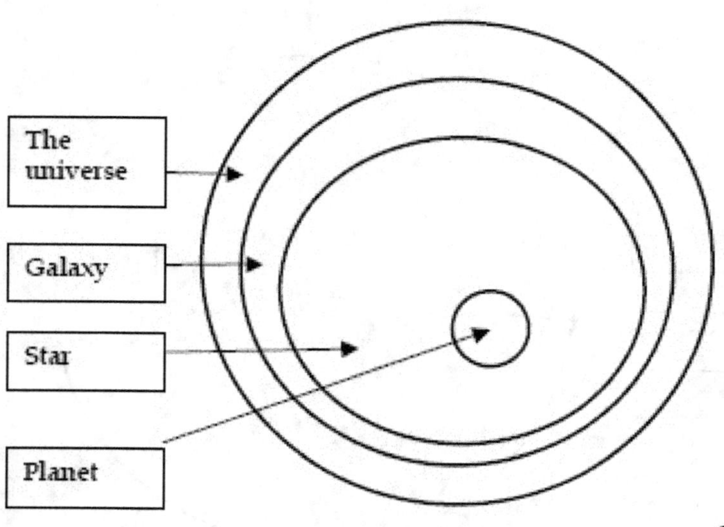
The universe
Galaxy
Star
Planet

?

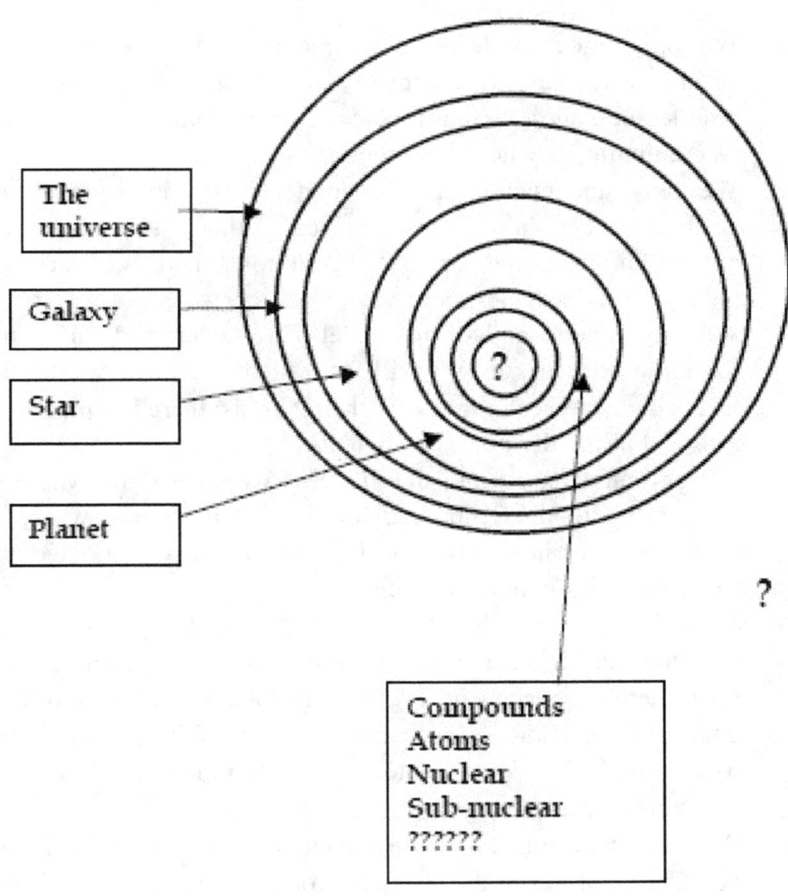

This is in essence a form of disguising what it is we observe as an inversion of '?'. The inversion factor simply inverts the left and the right as seen in a mirror.

In the case of '?' found 'outside' and '?' found 'inside' as seen through the observation of the tunnel of abstraction or what might more generically be called the tunnel of perception.

Volume 7: The Error of Boethius, detailed a few principles regarding the symmetry of abstractual and physical relationships:

The principle of symmetry:

1. We only understand half the model if we understand where seamlessness lies in multiplicity for if there is a place where seamlessness lies in multiplicity, then it holds there is a place where multiplicity lies in seamlessness.
2. We only understand half the model if we understand where abstraction lies in the physical for if there is a place where abstraction lies in the physical, then it holds there is a place where the physical lies in abstraction
3. We only understand half the model if we understand where divine foreknowledge lies in free will for if there is a place where divine foreknowledge lies in free will, then it holds: there is a place where free will lies in divine foreknowledge.
4. We only understand half the model if we understand where non-Centricism lies in Centricism for if there is a place where non-Centricism lies in Centricism, then it holds there is a place where Centricism lies in non-Centricism
5. We only understand half the model if we understand where omni benevolence lies in omniscience, omnipotence, and omnipresence for if there is a place where omni benevolence lies in omniscience, omnipotence, and omnipresence, then it holds there is a place where omniscience, omnipotence, and omnipresence lies in omni benevolence
6. We only understand half the model if we understand where the non-Cartesian lies in the Cartesian for if there is a place where the non-Cartesian lies in the Cartesian, then it holds there is a place where the Cartesian lies in the non-Cartesian. $d/t = v/1$
7. We only understand half the model if we understand where the Cartesian lies in the non-Cartesian for if there is a place where the Cartesian lies in the non-Cartesian, then it holds there is a place where the non-Cartesian lies in the Cartesian. $t/d = 1/v$
8. We only understand half the model if we understand where nothingness lies in the physical for if there is a place where

nothingness lies in the physical, then it holds there is a place where the physical lies in nothingness.

9. We only understand half the model if we understand where time is directly proportional to distance for if there is a place where time is directly proportional to distance, then it holds there is a place where the inverse of time is directly proportional to the inverse of distance. E/m = d/t
10. We only understand half the model if we understand where seamlessness lies in multiplicity for if there is a place where seamlessness lies in multiplicity, then it holds there is a place where multiplicity lies in seamlessness.

So where does symmetry lie?

Mathematically the demonstration of symmetry lies with the first mirror introduced by Hegel lies with the inversion of the Aristotelian perception that the functionality of time and the functionality of distance affect the observer.

Hegel inadvertently introduced the first mirror, introduced the inversion of the left becoming the right, with his expansion of Kant's idea that the observer affects the functionality of time.

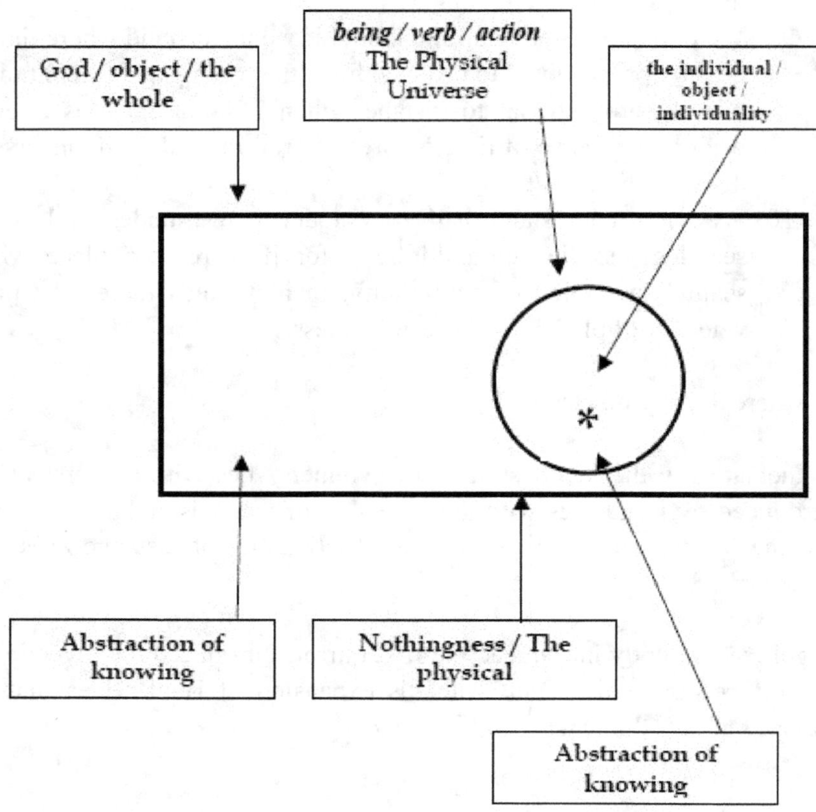

Opening the system up we obtain:

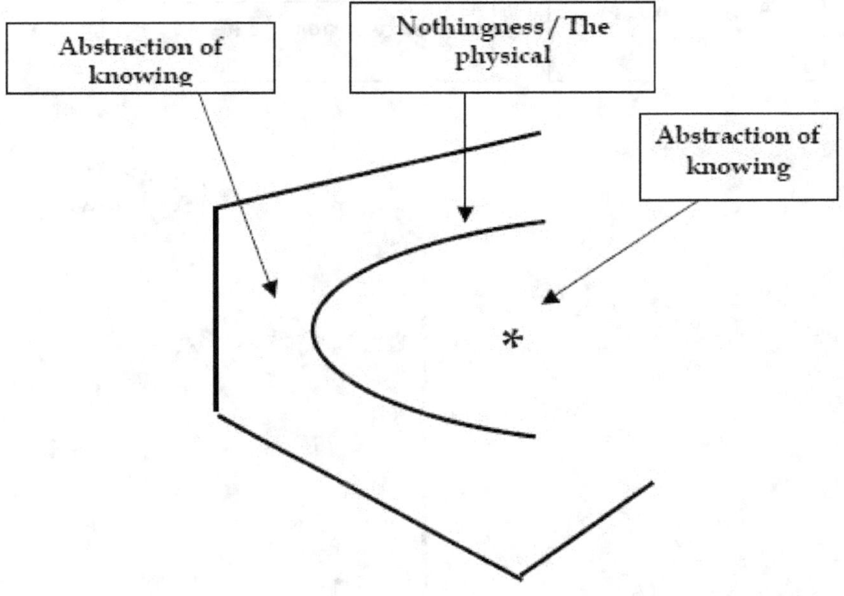

What appears to be is the concept of 'knowing' the verb and 'knowing' the noun, the knowing of knowing. Knowledge appears to be the universal building block of existence, metaphysically speaking. Straightening the lines even further we obtain:

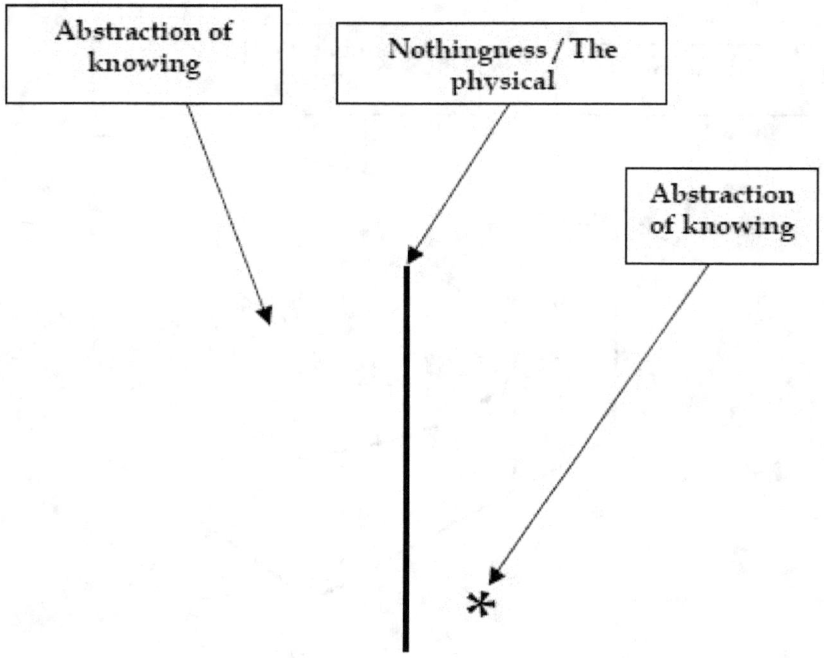

But what happened to the line segments forming the boundary of the whole?

The boundaries of the whole have been discarded to give the diagram its true appearance of existing in a limitless whole.

Or:

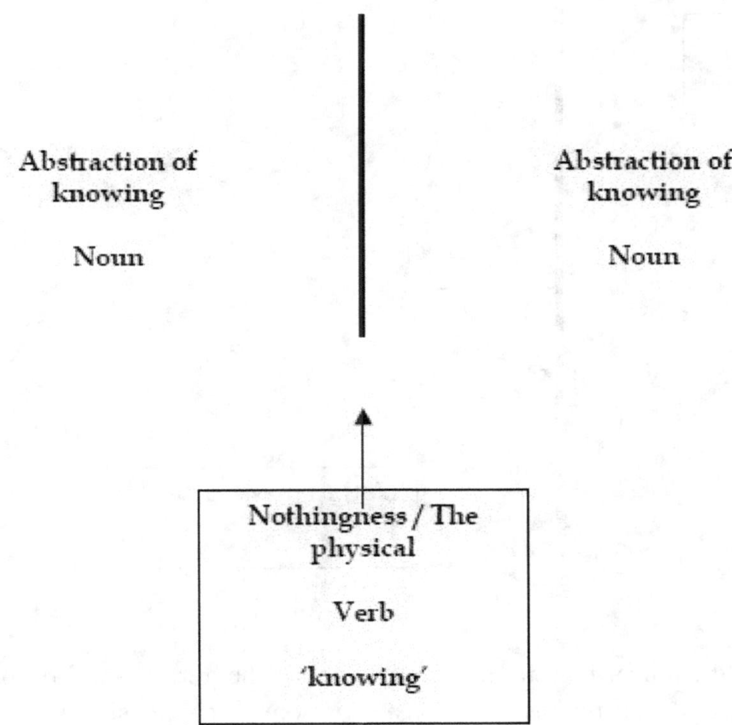
Abstraction of knowing
Noun
Abstraction of knowing
Noun
Nothingness / The physical
Verb
'knowing'

The left hand is the right hand and the right hand is the left hand. In terms of distance and time the mirror becomes

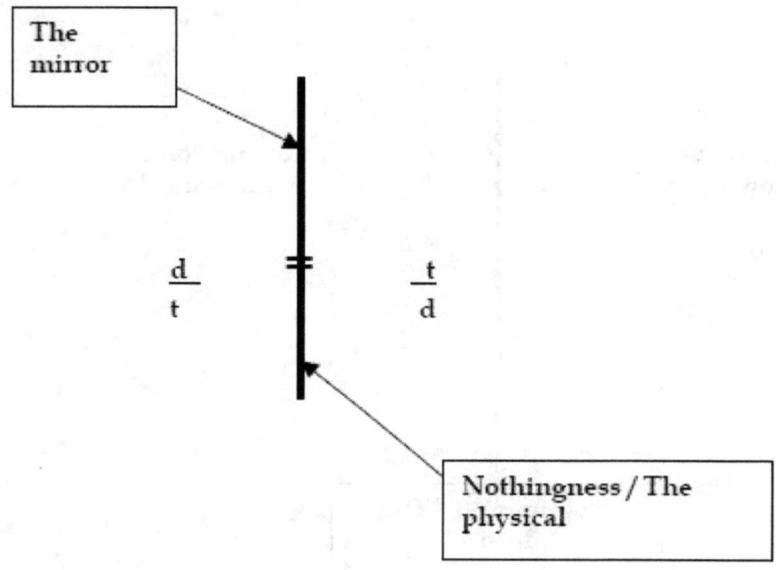

Why would the mirror be a form of nothingness? The mirror is a form of nothingness for the mirror has no dimension, is nonexistent physically.

Why would the mirror be physicalness, which implies the physical, is itself a form of nothingness?

The physical is limited in terms of existence and its presumed annihilation leads to the conclusion that the physical, although it may be reconstructed, is not permanent as ‘it was’ but rather is permanent in ‘it will be’.

The universe is a verb, action as opposed to being a noun, an object. A verb is not; rather a verb is simply a statement regarding the acknowledgement of the existence of an object.

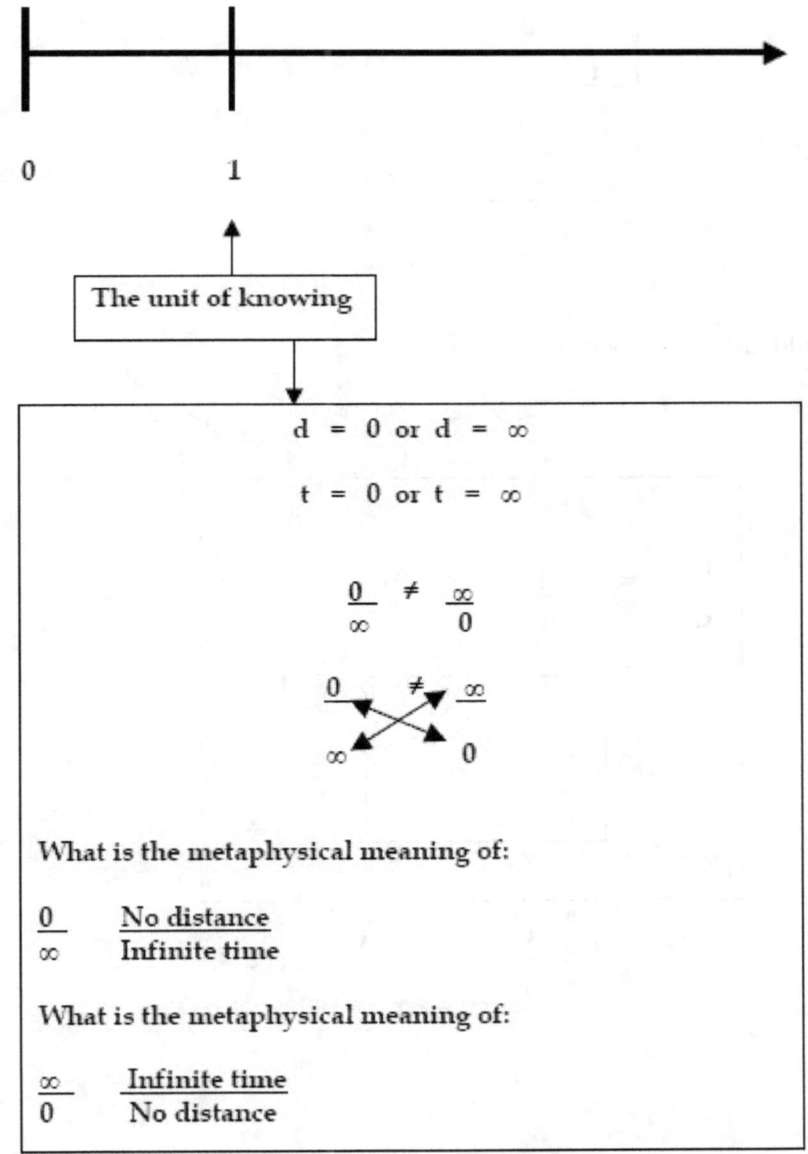

0
1
The unit of knowing
d = 0 or d = ∞
t = 0 or t = ∞
0/∞ ≠ ∞/0
What is the metaphysical meaning of:
0/∞ No distance / Infinite time
What is the metaphysical meaning of:
∞/0 Infinite time / No distance

For:

$$\frac{d}{t} = \frac{d}{t}$$

Could just as well be expressed as:

$$\frac{t}{d} = \frac{t}{d}$$

$t = 0 \text{ or } \infty$

$d = 0 \text{ or } \infty$

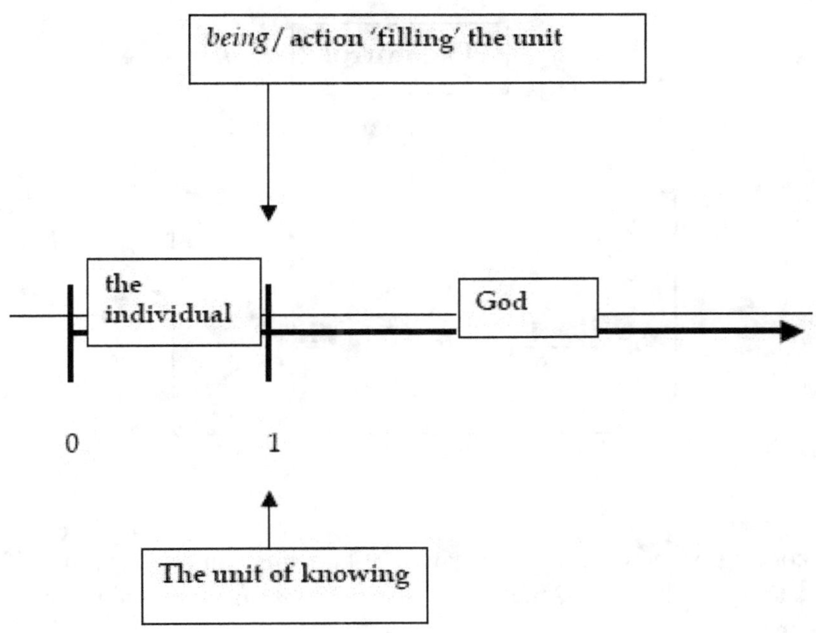

Why is there no 'left, symmetry to zero?

Zero is not the point of demarcation for knowing for there is no 'negative' to knowing.

Metaphysically, the point of symmetry for knowing is one not zero and thus the point of demarcation for knowing is one. In essence there is:

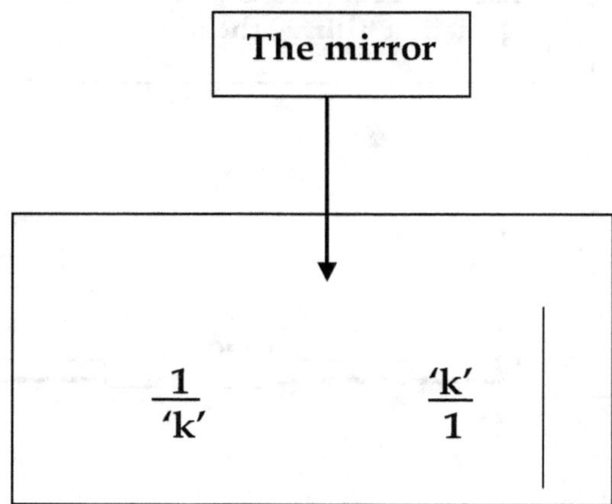

The concept of there being no negative of knowing can be exemplified by the statement: There is no 'negative' of air. There may be a void of air but not a 'negative' of air.

In terms of physicality:

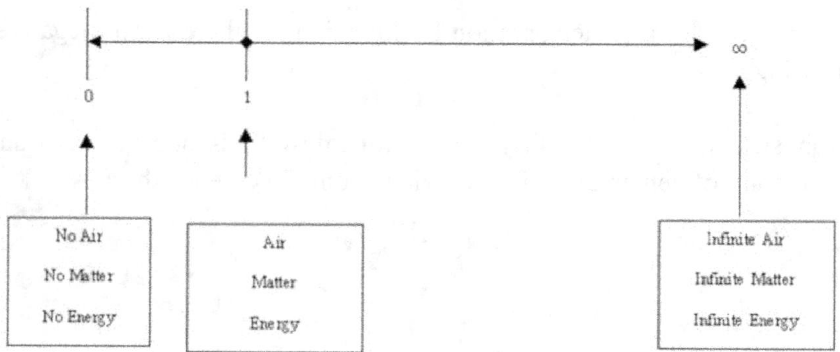

It is not:

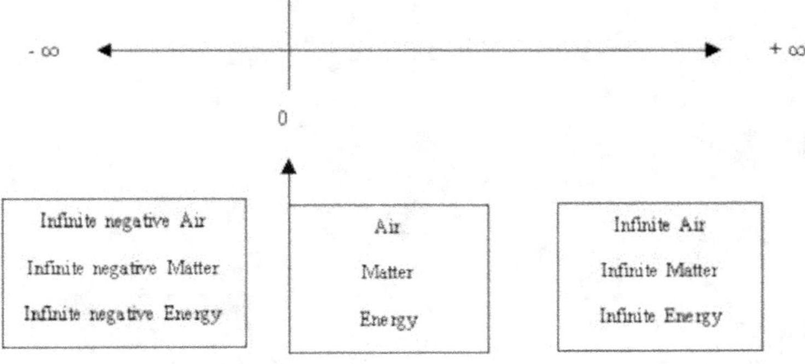

Likewise in terms of knowing, the noun, it is not:

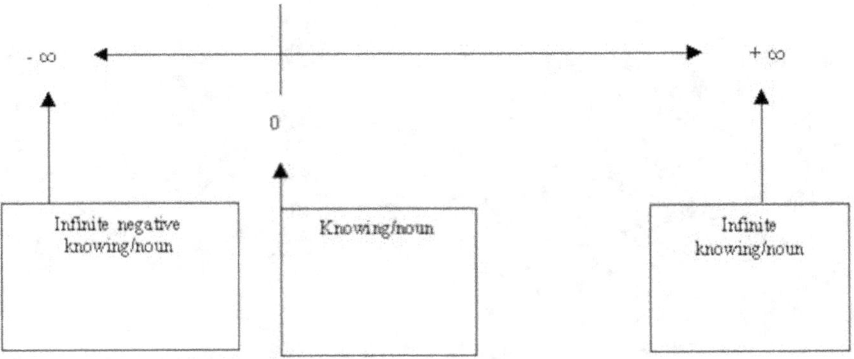

Rather it is:

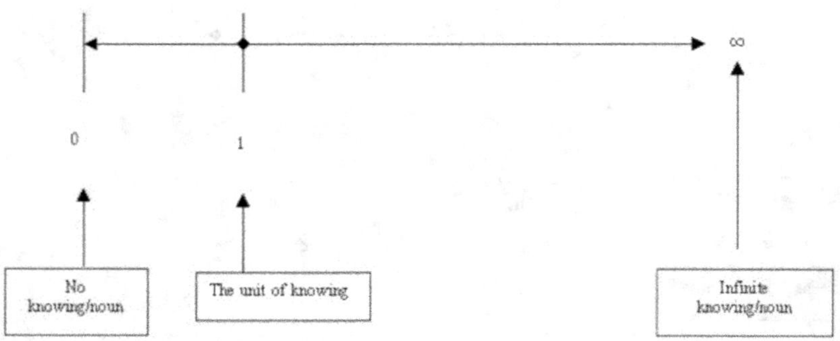

Where **0** and ∞ are approached rather than 'passed through'.

22. The tunnel of perception

We are now ready to revisit the tunnel of abstraction in the physical form.

To distinguish the physical form of the tunnel from the abstractual form of the tunnel we will refer to the physical form of the tunnel as the tunnel of perception since it is through the physical senses we do our observing and it is from our observing that we draw most of our understandings regarding abstractual perceptions.

Once having established our tunnel of perception, we will examine how the physical form is ‘attached’ to the tunnel of abstraction.

The tunnel of perception

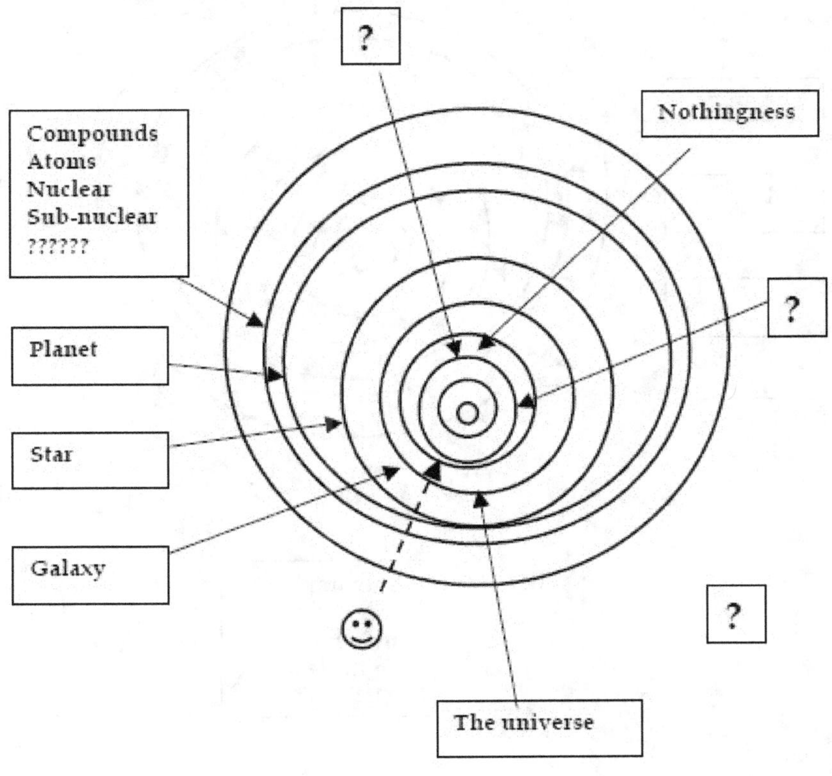

As we can see we have already created the basics for our tunnel earlier. A few modifications are in order:

In this tunnel, a second question mark appears beyond the physicalness of nothingness. The density of physical space becomes less and less the further 'out' we move. The question becomes: What is less dense than the physicalness of nothingness itself?

If we turn our tunnel around, we get:

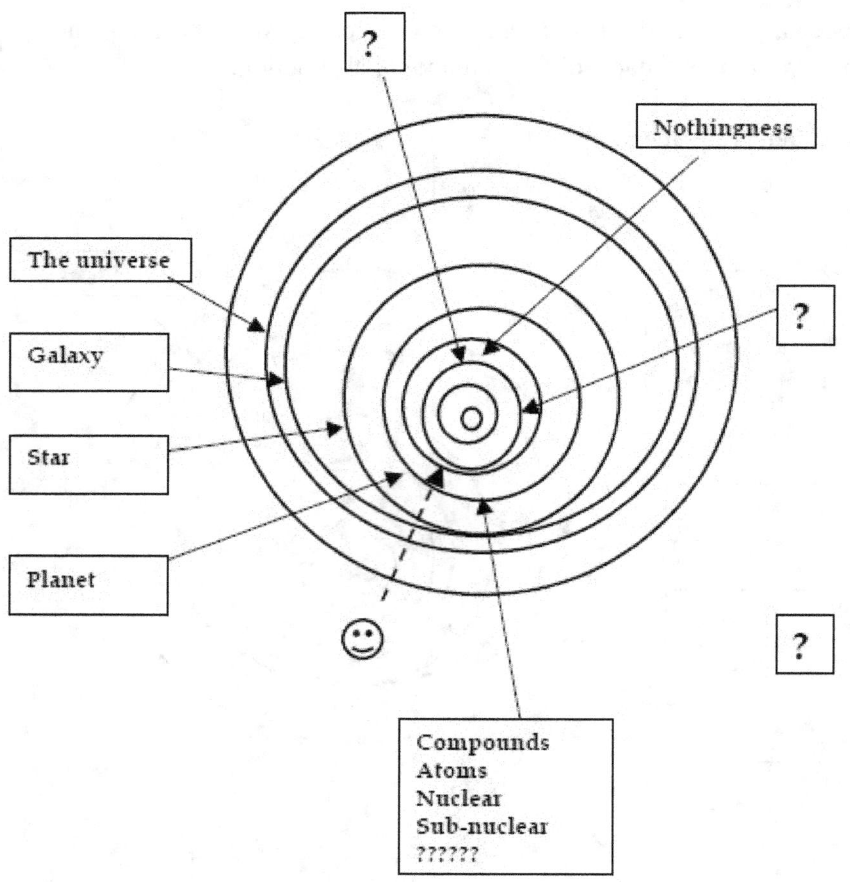

Again we see: In this tunnel, a second question mark appears beyond the physicalness of nothingness. The density of physical space becomes greater and greater the further 'in' we move. The question becomes: What is denser than the physicalness of nothingness itself?

Denser/less dense, it is the understanding of 'density' we 'gravitate' towards because density represents a characteristic of the physical and it is the understanding of the physical we think we seek because it is the physical with which we are most closely associated. In fact, however, the physical is not a 'thing' but rather reduces down to 'no' thing, reduces down to nothingness as we approach both ends of the tunnel of perception. The physical approaches the 'primal atom'($\infty / 1$) in one direction of the tunnel and approaches the universal building block ($1 / \infty$) in the other direction of the tunnel. This understanding leads us to Einstein's concepts of relativity and how they relate to 'i". The understanding regarding the relationship of 'i' to the physical leads us to understanding that time and distance is only variable where time is functional. Time and distance are only variable in the coherency of the constancy of the physical and time and distance are only constant in the incoherency of the variability of the abstract. To better understand such a concept it will help to use graphics.

We are now ready to expand our tunnel of perception to include our tunnel of abstraction:

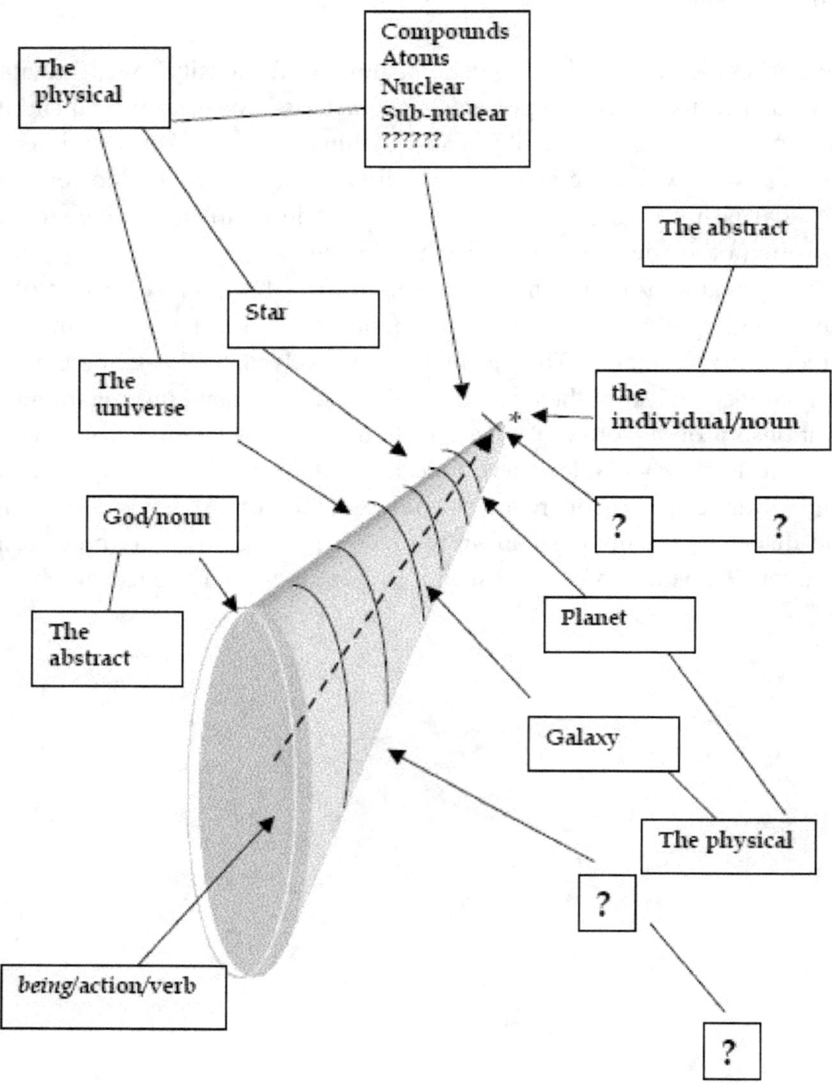

Having laid the ground work needed to understand the metaphysical significance of ‘i’, we are now ready to apply the fundamental ground work to the concept regarding what possible metaphysical meaning, the quotient, the square root of the distance divided by the square root of time, could possible have to the whole of reality versus physical reality.

Part IIb: The Einsteinian 'i': The Constant Variable Equals the Square Root of the Distance Divided by the Square Root of the Time

23. Introduction

We have examined the mathematical and metaphysical relationship regarding the abstract as it applies to individual numbers both imaginary numbers and real numbers. We have examined Zeno's work as it applies to the separate existence of real/Cartesianism and the 'real illusion'/ 'imaginary/non-Cartesianism.

We have examined the mathematical relationship regarding inverse relationships as it applies to the metaphysical linear relationship regarding imaginary and real numbers and we have examined Hegel's work as it applies to the separate existence of passive observation/Cartesianism and active observation/non-Cartesianism. Mathematically we observed this relationship emerging from the concept: the inverse of time being equal to the inverse of distance.

Now we are ready to examine the mathematical relationship regarding relativity as it applies to the metaphysical quadratic relationship regarding imaginary numbers and real numbers which is implied by the metaphysical system of the individual acting within God/panentheism.

In short we are ready to examine the metaphysical relationship of imaginary numbers and real numbers and how it is this mathematical/scientific understanding supports the concept regarding the metaphysical existence regarding the Cartesianism and the non-Cartesianism existing as simultaneous independent states of action found within the whole.

Mathematically we will observe this relationship emerging from the concept: the square root of the speed of light squared (velocity squared) found within the equation: E = mc(2).

We have now reached the point from which we can begin to understand what it is Einstein inadvertently had to offer us metaphysically which we were unable to 'see' before Einstein developed his perceptions of physics and before the new perception of the individual acting within God had been put into place.

We are on our way towards expanding our understanding regarding the concept regarding the functionality of concrete variability of time as opposed to abstract constancy of time.

In this last portion of the volume, it is not the absolute values of energy, matter, distance, or time upon which metaphysics relies to expand its past models, rather it is the presence of 'i' and the lack of the presence of 'i' which become the focus point of metaphysics.

It is our ability to understand the relationship existing between time, distance, matter, and energy upon which metaphysics must explore.

What is to emerge is the understanding that positive and negative values in terms of both the 'real'/concrete and the 'real illusion'/ 'imaginary'/ abstract have a part to play in our understanding the metaphysical relationship existing between 'i' and the lack of 'i'.

It is the ability of the model, the individual acting within God, the non-Cartesian powered by the Cartesian, which reinforces the validity of the simultaneous integrated existence of the Cartesian and the non-Cartesian, the individual acting within God, as opposed to either simple Cartesianism or simple non-Cartesianism being the model of reality.

In addition, it is the inability of either the Cartesian or the non-Cartesian to explain such a relationship which further reinforces the validity of the simultaneous integrated existence of Cartesian and the non-Cartesian, the individual acting within God, as opposed to either simple Cartesianism or simple non-Cartesianism being the model of reality.

24. The square root of Einstein's equations: 'i'

#1

E = mc2

E - mc2 = 0

(E - mc2) - 0 = ?

#2

E = mc2

E/m = c2

√(E/m) = √(c2)

$$\frac{\sqrt{E}}{\sqrt{m}} = \sqrt{(c2)}$$

Because we are not interested in the numerical 'value' of the square root of E, m, and c we will treat the variable designation of E, m, and c as forms of constants be it constant variables, variable constants, or constant consistencies of change. As such we obtain:

$$\frac{+/- \ E(1/2)}{+/- \ m(1/2)} = +/- \ c$$

becomes simply:

$$\frac{+/\_ \ E}{+/- \ m} = +/- \ c$$

The mathematical 'correctness' of such a statement is not of issue here. The process allows us to simplify what it is we are attempting to understand metaphysically. With the mathematical issue stated we can now move to four scenarios the equation implies:

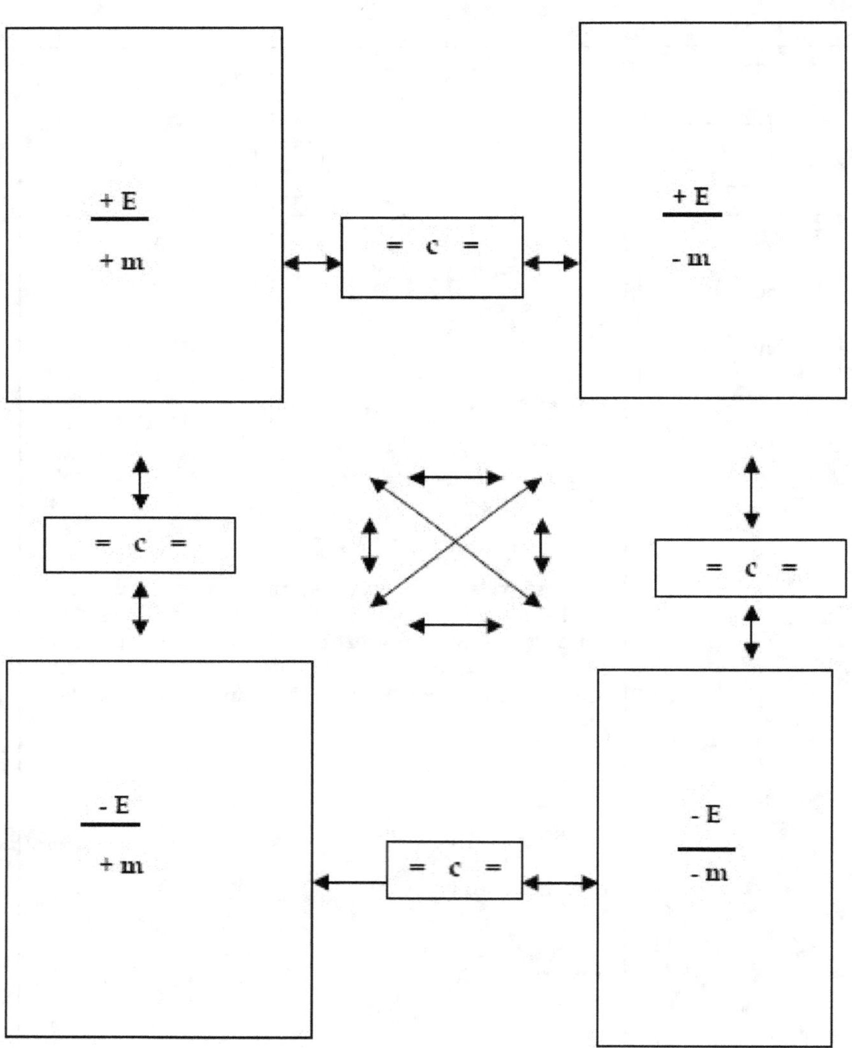
+ E
+ m
= c =
+ E
- m
= c =
= c =
- E
+ m
= c =
- E
- m

And secondly:

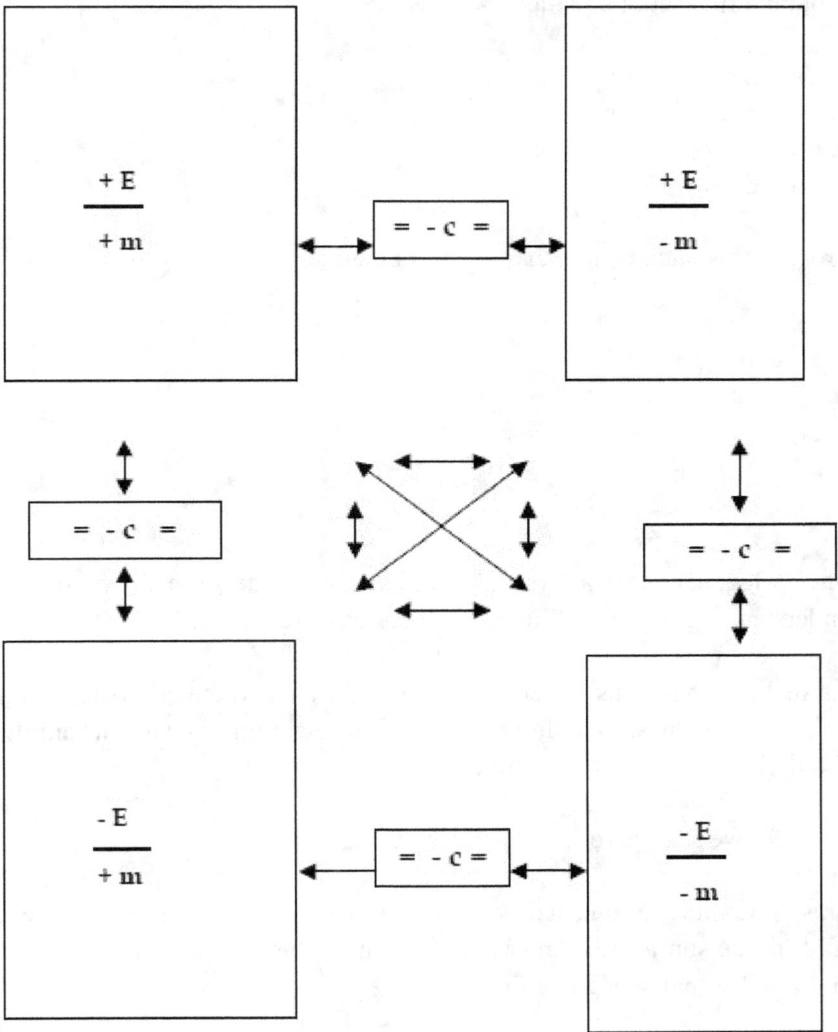

Regardless of the equations only two results emerge if one places the concepts of energy and matter in the realm of constancy. The two emergent results become the positive of the speed of light or the negative of the speed of light:

The schematic gives us eight options.

From the first set of equations we obtain:

1. c = E/m
2. c = - E/m
3. c = E/ - m
4. c = - E/ - m

From the second set of equations we obtain:

5. - c = E/m
6. - c = - E/m
7. - c = E/ - m
8. - c = - E/ - m

The Hegelian metaphysical system was developed before the understanding regarding Einstein's theories of relativity.

As such, we will consider both the speed of light and the quotient of E/m to be absolute values. In addition we will consider E/m to be a constant. E/m, should it exist would be a 'universal constant'.

As such, we shall give it a unit value of one.

This is no more an unusual process than designating the distance from the earth to the sun to be a unit value, the unit value of one astronomical unit or 1 au. We shall designate E/m as 1

This is not to say E/m is one nor is it saying E/m is a constant. Rather E/m is considered to be a variable constant of physicality.

The result is that all eight options when the absolute value is applied become simply:

$$c = 1$$

The speed of light is the ratio of distance to time. This gives us:

$$d/t = 1$$

or

$$1 = d/t$$

In terms of metaphysic, what does the number one imply and what does the ratio of time to distance imply?

The number one implies 'a' location where time and distance can be found 'within' which change can originate.

The number of such locations has no meaning 'within' timelessness in terms of an absolute value. However, as perceived by awareness immersed within a universal fabric of time,

$$\sum_{1}^{\infty}$$

represents the potential 'number' of locations/universes which could exist within timelessness within which change of the timeless could be generated.

In terms of the symbolic representation of d/t, the quantity, d/t represents velocity. Velocity is the relationship between two abstractual concepts, distance and time, generated by the physical.

$v = d/t$

Where $v = c = +/- 1$

$1 = d/t$

or

$1/1 = d/t$

or

$1/1 = t/d$

The point is:

$d = t$

What does distance equals time imply? Within the purely physical, such a concept is meaningless. Within the metaphysical, the direct relationship is very significant.

Metaphysically the absolute value of distance being equal to the absolute value of time implies that neither time nor distance is more important than the other.

Relative value is meaningless within the realm of the abstract. Abstractual concepts have no relative value one to another because the concept of a universal fabric of time and distance is absent.

The universal fabric by which we 'compare' the value of one to another is non-existent as a 'universal' marker within the purity of the whole of the abstract 'found' to exist 'outside' the physical**...**

There is more than just the elimination of 'relative value' implied in the concept: d = t which needs to be examined.

25. Einstein's mirror revisited

Hegel introduced the first metaphysical mirror.

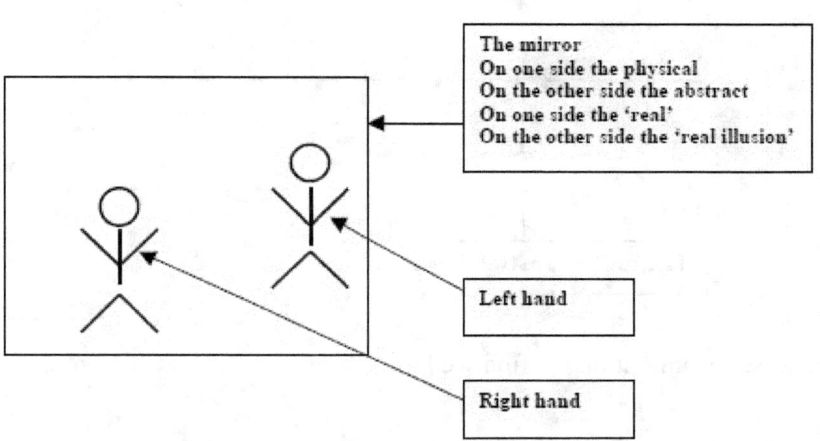

Einstein inadvertently introduces the second metaphysical mirror. The progression emerges as follows:

In physicality we seek the multiplicity of parts from the smallest to the largest:

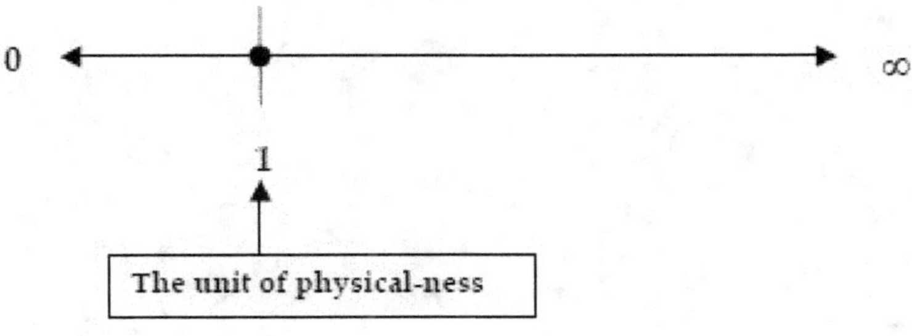

In seamlessness we must seek the mirror image of physical-ness not the opposite. As such we must reverse the direction and seek one-ness/singularity:

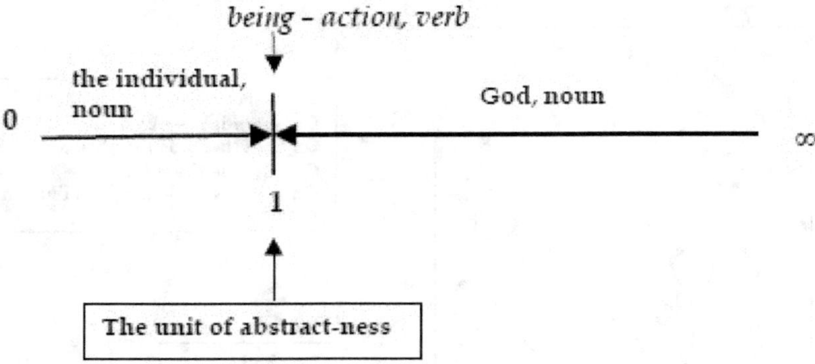

On closer examination we find we have:

this becomes:

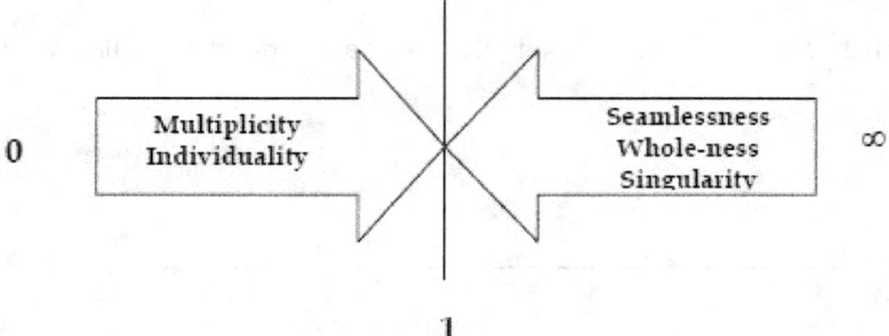

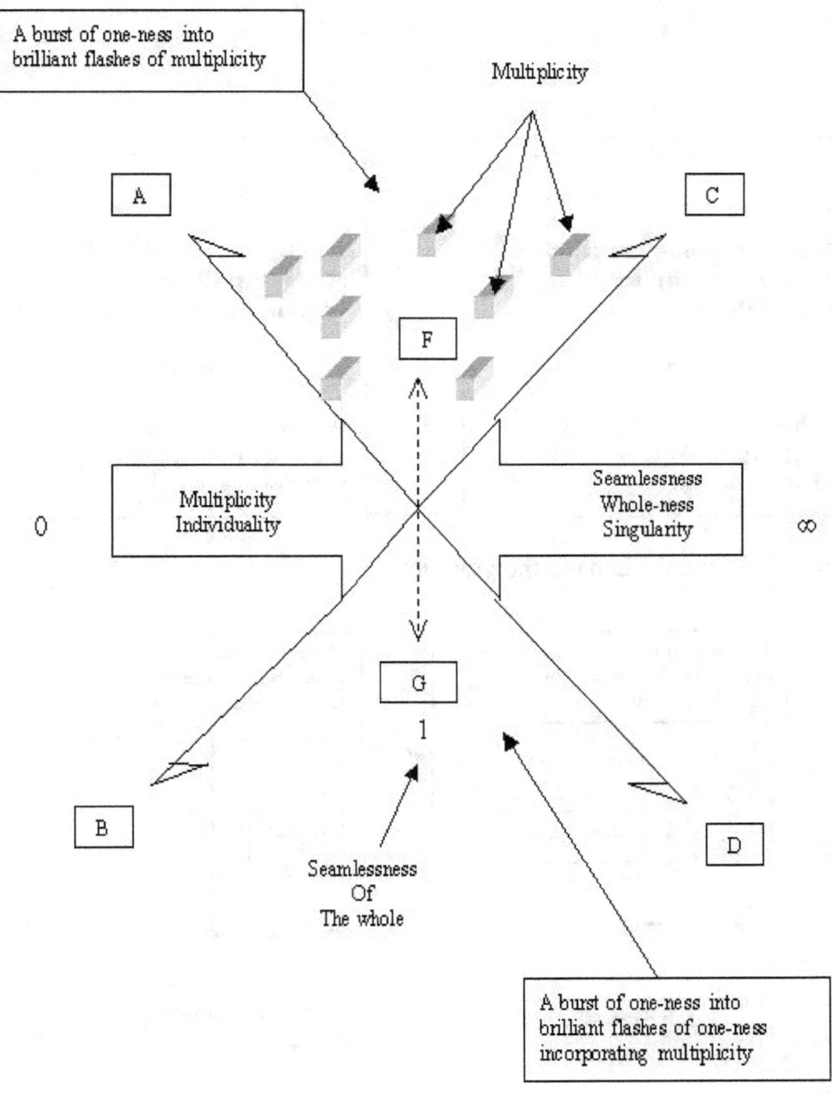

Curve the plane of the paper and point A meets point B and point C meets point D forming a rectangular - God, and point F meets point G forming a circle of physical-ness - the individual, through action of unity - *being*.

Symbolically we obtain:

$\frac{v'}{v} = \frac{t}{d}$ **Where v = one / the whole of the real / multiplicity / physical-ness** **and** **Where v′ = the whole of the 'real illusion' / singularity / abstract-ness**	**and**	$\frac{v}{v'} = \frac{d}{t}$ **Where v′ = one / the whole of the 'real illusion' / singularity / abstract-ness** **and** **Where v = the whole of the real / multiplicity / physical-ness**

From the above we expand the graphic:

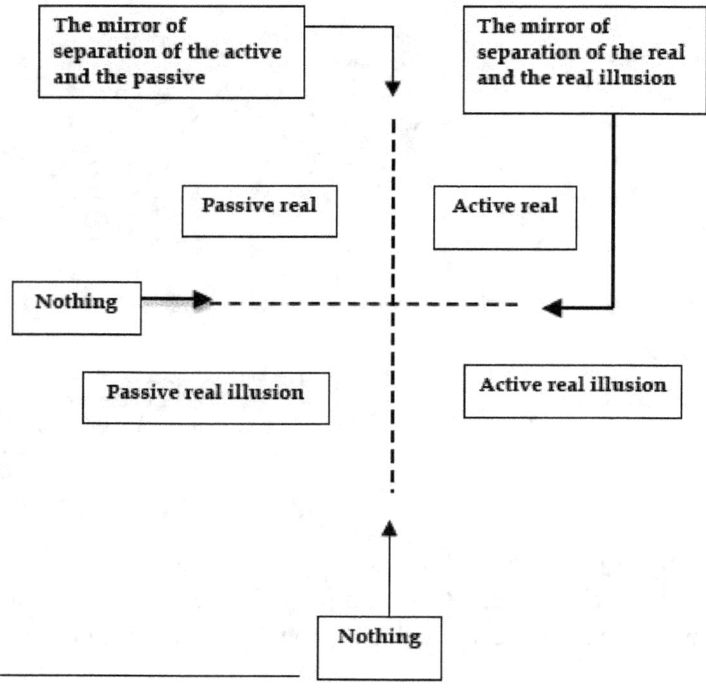

Scientifically the mirrors are representative of Aristotle and Einstein:

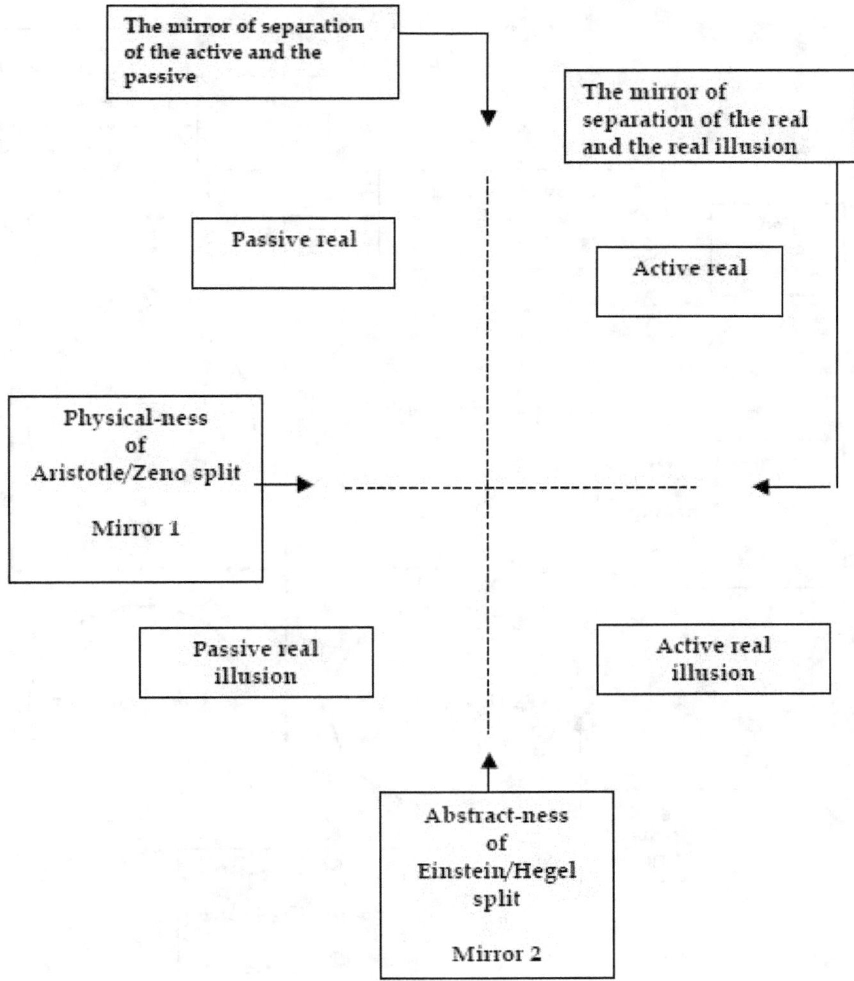

Expanding metaphysically upon our diagram we obtain:

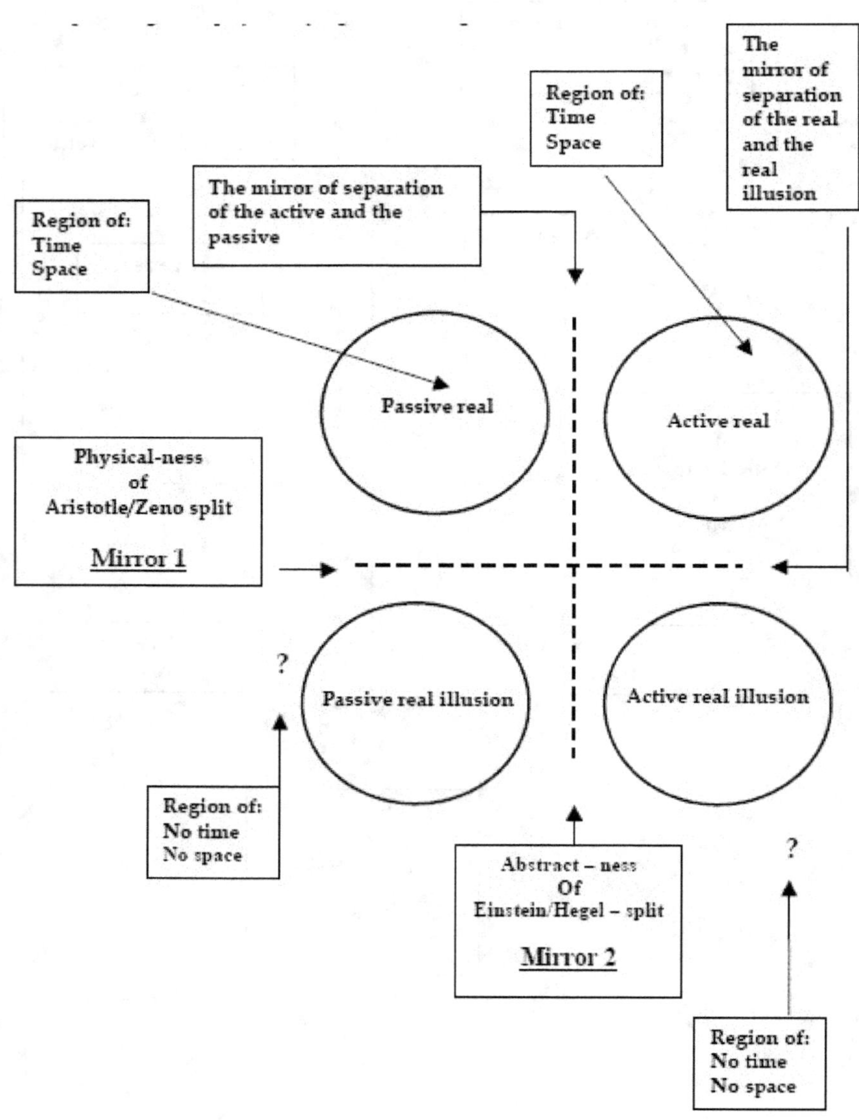

Further metaphysical expansion gives us:

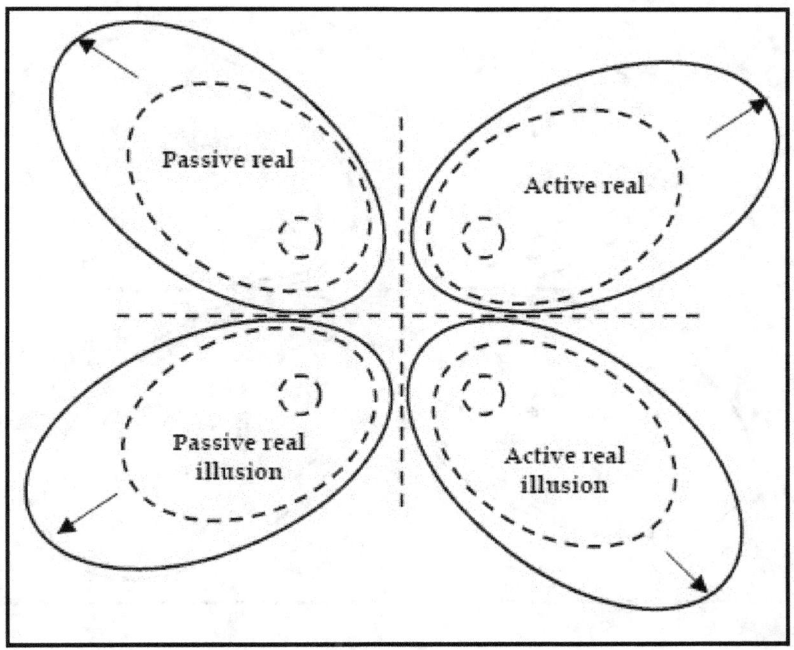

Infinite growth 'outward' now possible since growth moves 'outward' into the void of space/distance and 'outward' into the void of time

And

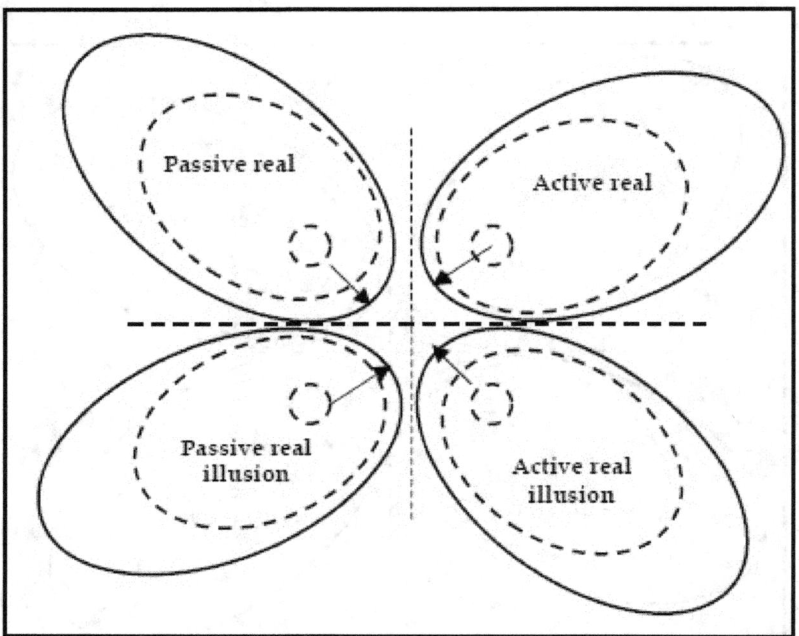

Infinite growth 'inward' now possible since 'growth moves 'inward' into the void of space/distance and 'inward' into the void of time

The graphic can be simplified as:

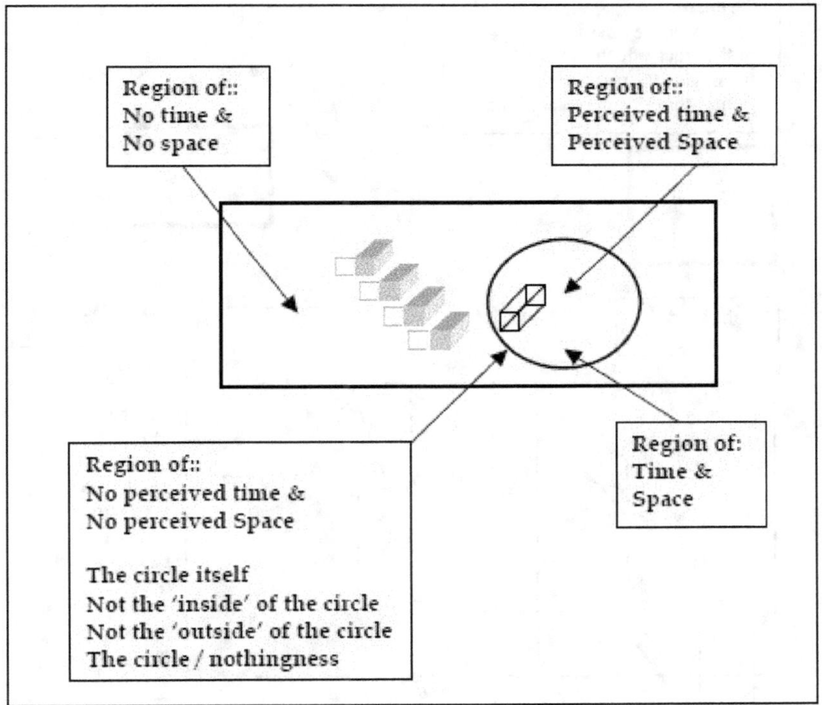

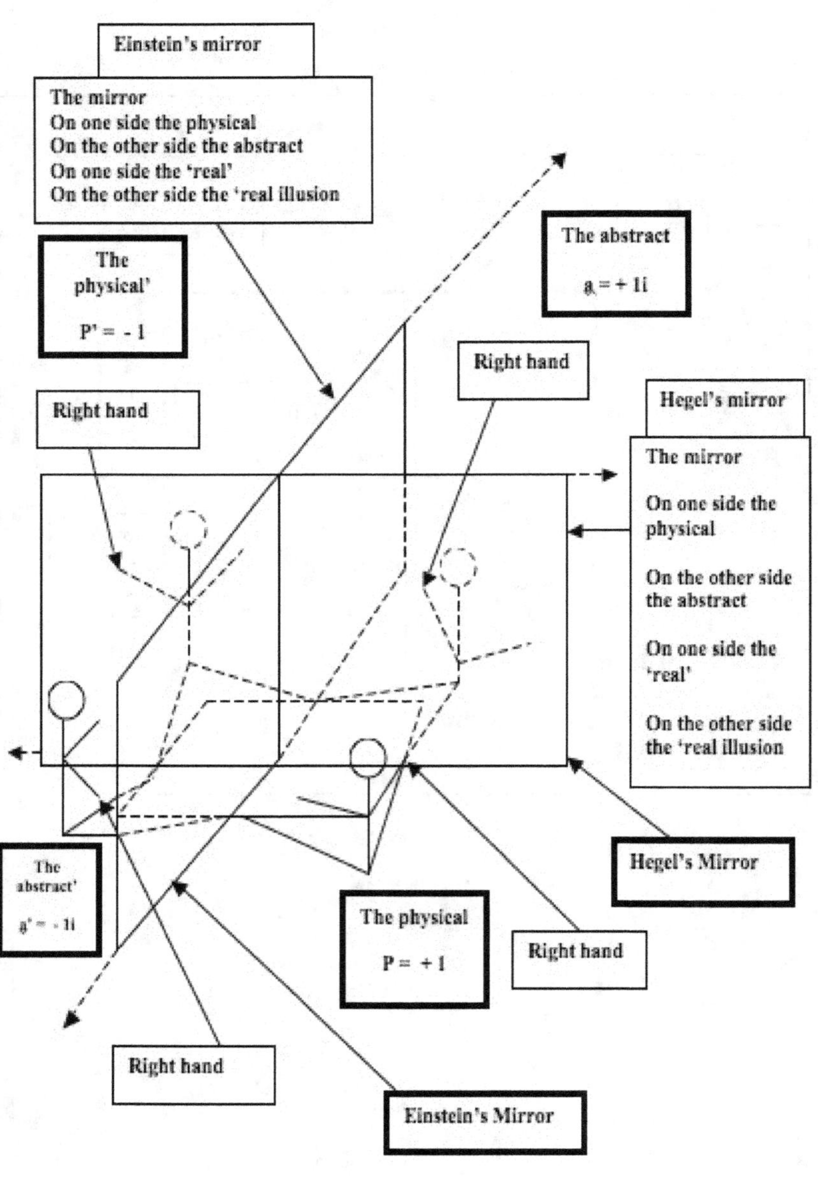
Einstein's mirror
The mirror
On one side the physical
On the other side the abstract
On one side the 'real'
On the other side the 'real illusion
The physical'
P' = - 1
The abstract
a = + 1i
Right hand
Right hand
Hegel's mirror
The mirror
On one side the physical
On the other side the abstract
On one side the 'real'
On the other side the 'real illusion
The abstract'
a' = - 1i
The physical
P = + 1
Hegel's Mirror
Right hand
Right hand
Einstein's Mirror

Now we have:

$$p = +1$$

$$a = 1i$$

$$p' = -1$$

$$a' = -1i$$

From this we can see four regions of existence: p, a, p’, a’.

26. Illusion

From Volume 10: The Error of Kant we observed:

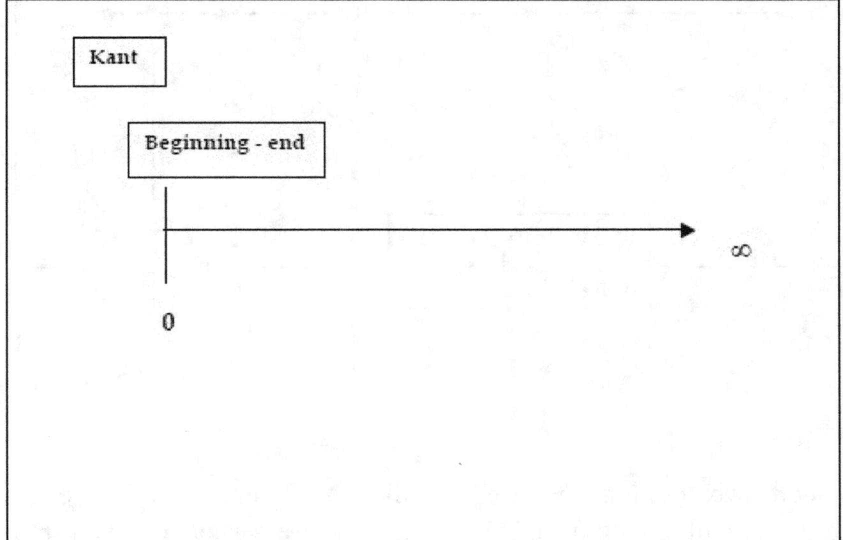

In Volume 11: The Error of Hegel, we then observed the expansion of the idea becoming:

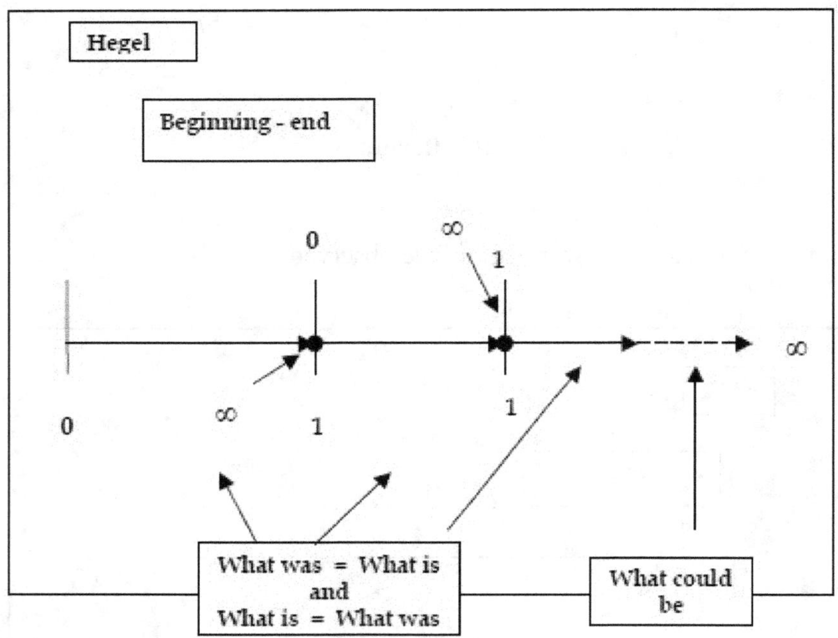

In short, two does not exist; we just call the next unit: Two. What exists is two ones, multiplicity of uniqueness exists. In essence we obtain infinite infinities.

Metaphysically we observe ‘what was’ as being identical to ‘what is’ and ‘what is’ as being identical to ‘what was’. ‘What will be’ is without time, simply is, and as such is a part of ‘what is’ and ‘what was’. ‘

What could be', however, 'is not', 'was not', and may not be. As such 'what could be' is unpredictable in all manners regarding what the term 'unpredictable' implies.

Metaphysically knowing action, *being*, is not 'set' but rather 'is determined' by the very action of knowing beings reacting to the unpredictability of random physical actions found within the physical, found within the realm of energy, matter, functional space, and functional time.

Heidegger stated: 'Why are there essents rather than nothing?'

Kant's work implies the question should be: If there is no nothing why is there essent/essents? If nothing does not exist then the '?' is the knowable and the unknowable:

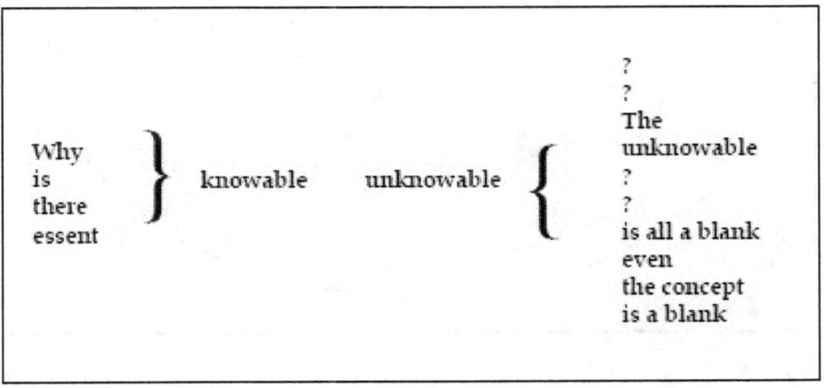

Again graphics help us understand the meaning regarding the above statement regarding abstraction:

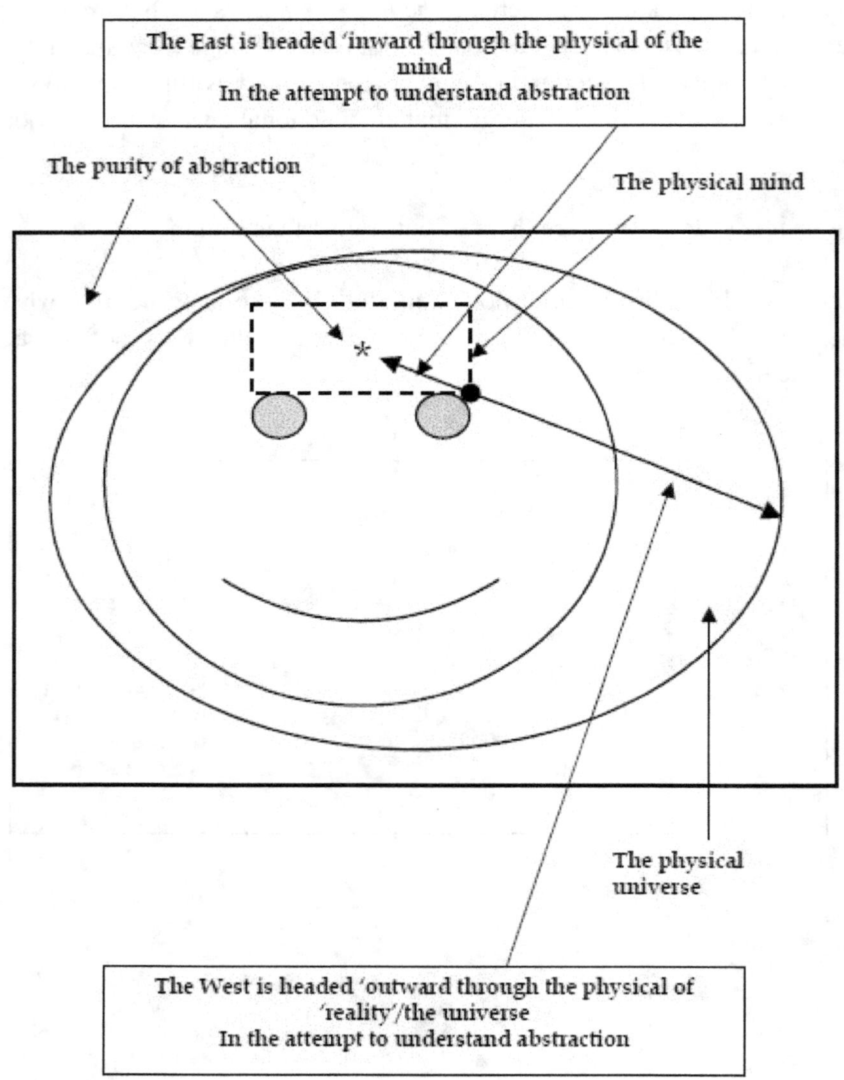
The East is headed 'inward through the physical of the mind
In the attempt to understand abstraction
The purity of abstraction
The physical mind
*
The physical universe
The West is headed 'outward through the physical of 'reality'/the universe
In the attempt to understand abstraction

The East looks inward to gain a sense of the individual unique whole part, namely understanding one's self and the relationship of one's self to the whole.

The West looks outward to gain a sense of the individual unique whole of the oneness of it all and the relationship of the whole to one's self.

Metaphysically, both fall short of their goal and will continue to fall short of their goal since neither is looking to understand the complete relationship of the whole/singularity to its/the parts/multiplicity and the parts/multiplicity to the/its whole/singularity.

It is this two-way interrelationship which mathematics/Einstein's equations can help metaphysics, the East, and the West understand.

The equation v = d/t, divided by time, t, being directly proportional to distance, d, is not specifically a metaphysical issue but the mathematical complexities of such a relationship can help us understand the most rational metaphysical characteristics of the whole, reality in total.

v = d/t

vt = d

Where v = The coefficient of constancy
The coefficient of constancy =
k

k = 1

1t = d

t = d

Likewise:

$1/v = t/d$

$d/v = t$

Where v = The coefficient of constancy
The coefficient of constancy = k
k = 1

$d/1 = t$

$d = t$

Aristotelian Cartesian metaphysical systems perceive the system to be closed and closed systems to incorporate infinity through the process of moving internally versus an open system which incorporates infinity through the process of moving outwardly.

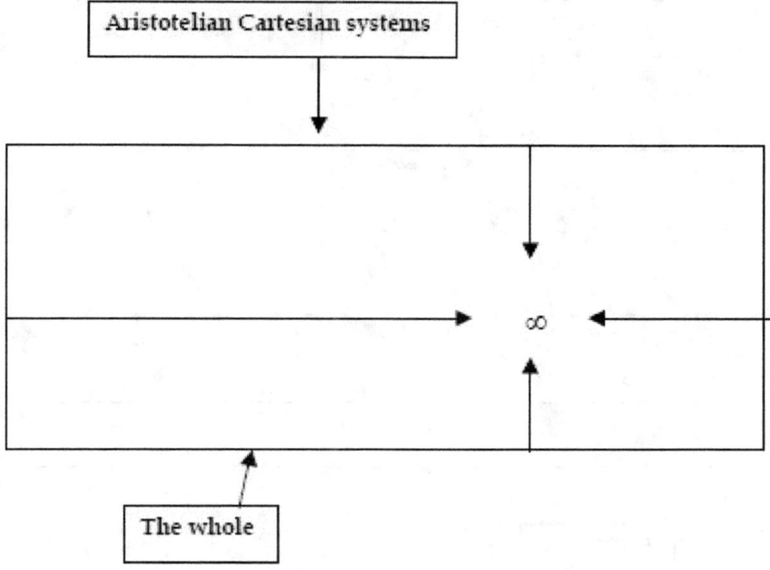

The system is a form of stagnation through internal expansion. The decay becomes apparent when one expands the graphic:

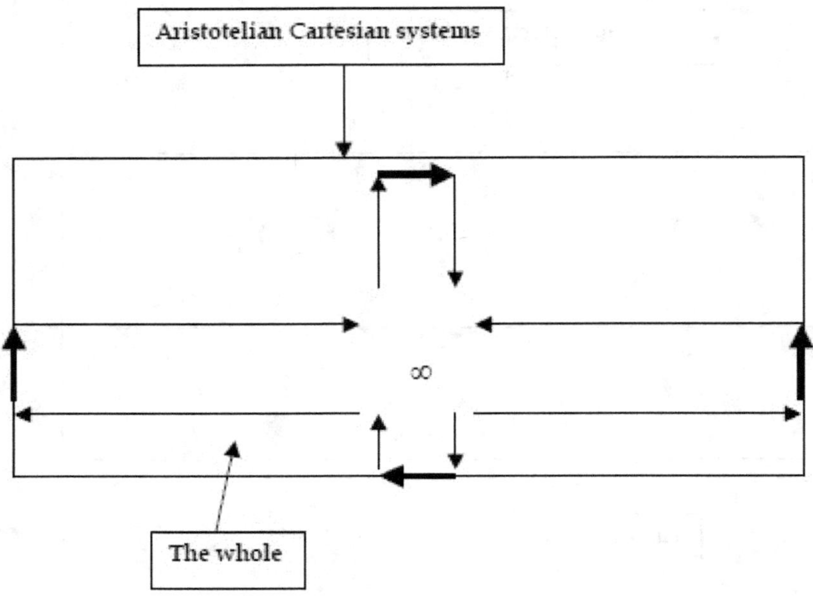

Infinity begins at the infinitely small and ends at the infinitely small. Multiplicity of parts exists but the whole remains what it is the whole with no potential to expand itself.

This is a one of the most insidious forms of stagnation for it is a form of stagnation concealed within the false perception regarding the insignificance of the infinite parts/multiplicity which ironically generates perceived insignificance for the parts of the whole or what might better be expressed as nihilism regarding the parts.

The second option is the non-Cartesian system: the outward expansion limited by its lack of inward growth.

Kant/Hegel non-Cartesian metaphysical systems perceive the system to be open and open systems to incorporate infinity through the process of moving externally versus closed system, which incorporate infinity through the process of moving inwardly.

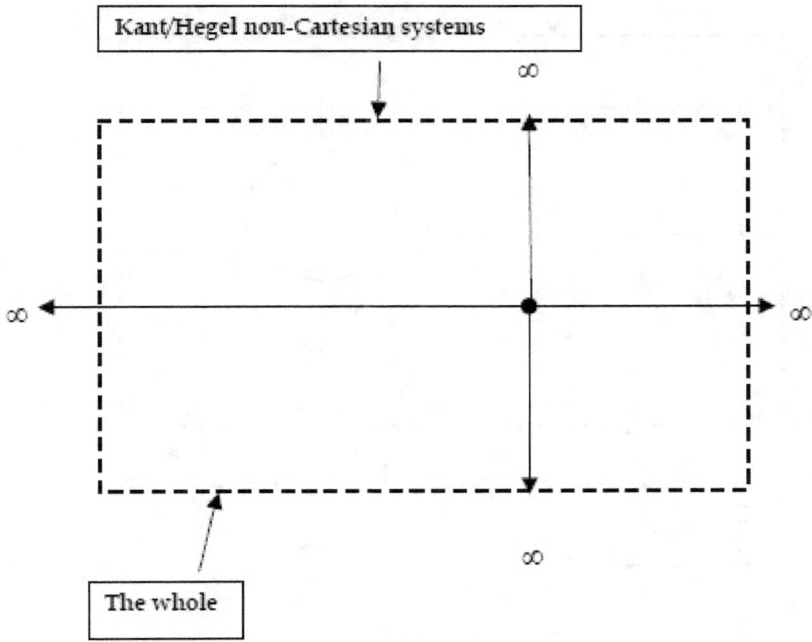

The system is once again a form of stagnation through external expansion. The decay becomes apparent when one expands the graphic:

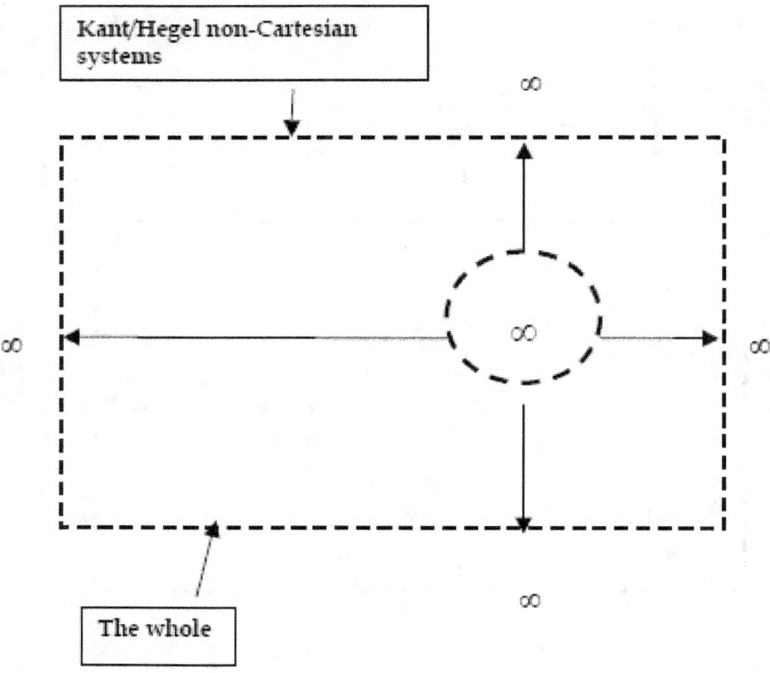

Infinity begins at the infinitely large and ends at the infinitely large. Singularity exists but the whole remains what it is the whole with no potential to expand its understanding of its infinite incremental parts.

This is a second insidious form of stagnation for this is a form of stagnation concealed within the false perception regarding the insignificance of the whole which ironically once again generates perceived insignificance for the parts of the whole or what might better be expressed as nihilism regarding the parts.

Both systems impact the parts of the whole adversely for each system generates perceptions of insignificance of the parts of the whole.

The significance of such generated perceptions becomes apparent to the parts themselves and to ourselves in particular since we are a part of the system itself.

Incorporating both systems, a non-Cartesian system powered by a Cartesian system provides the infinite expansion in both the inward and outward directions, which allows for the growth of the parts and the growth of the whole. Such a system, the individual acting within God, generates:

---

1. Neither a stable whole nor stable parts
2. Neither a decaying whole nor decaying parts

Rather the system of the individual acting within God generates:

A growing dynamic whole and growing dynamic parts.

The illusion then becomes the perception that one or the other is the system. The illusion then becomes the perception that either the system is a form of closed system, Cartesian, or the system is a form of open system, non-Cartesian.

The reality of the scenario lies not in the Cartesian nor in the non-Cartesian but rather in both the Cartesian and the non-Cartesian. The two exist simultaneously independent yet simultaneously dependent one upon the other.

Regarding the Cartesian and the non-Cartesian: The concrete/the physical perceives itself/ the Cartesian, to be real and the non-Cartesian/the abstract to be an illusion. Simultaneously the abstract/ the non-Cartesian perceive itself to be real and the Cartesian/the physical to be an illusion.

What actually evolves as the scenario is: One is the real illusion when the other is the real. Which is which depends upon 'where' it is one does the perceiving, is dependent upon one's relative position of perception.

How then does one get to the 'real illusion' from the mathematical Aristotelian perception of : v = d/t?

Within the equation, v = d / t, we will take 'v' to be a coefficient of constancy. With the coefficiency of constancy being '1' we obtain the real and the real illusion:

---

27. The real and the real Illusions

+ E	=	energy
- E	=	anti-energy
+ m	=	matter
- m	=	anti-matter
+ c	=	the speed of light in the vacuum of the physical: velocity = physical distance ÷ physical time
- c	=	the speed of light in the vacuum of the abstract: velocity = abstract distance ÷ abstract time

E c(2)	= m
E ÷ m	= c(2)
√E ÷ √m	= √c(2)
± E(1/2) ÷ ± m(1/2)	= ± c

- E c(2)	= - m
- E ÷ - m	= c(2)
√(- E) ÷ √(- m)	= √c(2)
± iE(1/2) ÷ ± im(1/2)	= ± c

There are two metaphysical questions which arise from the two graphics:

First: The real

If energy and matter are equated to the speed of light in a vacuum, what could this possibly mean metaphysically?

The two graphics demonstrate the validity of the metaphysical system of the individual acting within God for Einstein's equation clearly demonstrated the relationship which exists between the abstract and the physical. The statements:

E	= m c(2)
E ÷ m	= c(2)
√E ÷ √m	= √c(2)
± E(1/2) ÷ ± m(1/2)	= ± c

- E	= - m c(2)
- E ÷ - m	= c(2)
√(- E) ÷ √(- m)	= √c(2)
± iE(1/2) ÷ ± im(1/2)	= ± c

Expands to become:

E	= m c(2)
E ÷ m	= c(2)
√E ÷ √m	= √c(2)
± E(1/2) ÷ ± m(1/2)	= ± d / +/- t

- E	= - m c(2)
- E ÷ - m	= c(2)
√(- E) ÷ √(- m)	= √c(2)
± iE(1/2) ÷ ± im(1/2)	= ± d / +/- t

Second: The 'real' illusion

E = m c(2)	- E = - m c(2)
E ÷ m = c(2)	- E ÷ - m = c(2)
√E ÷ √m = √c(2)	√(- E) ÷ √(- m) = √c(2)
± E(1/2) ÷ ± m(1/2) = ± c	± iE(1/2) ÷ ± im(1/2) = ± c
Then if c is velocity and if velocity is the coefficient of constancy k which equals 1:	Then if c is velocity and if velocity is the coefficient of constancy k which equals 1:
± E(1/2) ÷ ± m(1/2) = ± k	± iE(1/2) ÷ ± im(1/2) = ± k
± E(1/2) = ± m(1/2)	± iE(1/2) = ± im(1/2)
Since we are not interested in the cardinal value of E and m we can ignore the factor of 1/2 and thus we can concentrate upon the positive and negative metaphysical significance of existence, concentrate upon the four scenarios:	Since we are not interested in the cardinal value of E and m we can ignore the factor of 1/2 and thus we can concentrate upon the positive and negative metaphysical significance of existence, concentrate upon the four scenarios:
+ E = + m	+ iE = + im
+ E = - m	+ iE = - im
- E = + m	- iE = + im
- E = - m	- iE = - im

If the opposite of energy (- E)and the opposite of matter (- m) are equated to the speed of light in a vacuum, what could this possibly mean metaphysically?

The two graphics demonstrate the validity of the metaphysical system of the individual acting within God for Einstein's equation inadvertently demonstrates the relationship which exists between the abstract and the physical. The statement:

- E	**= - m c(2)**
- E ÷ - m	**= c(2)**
√(- E) ÷ √(- m)	**= √c(2)**
± iE(1/2) ÷ ± im(1/2)	**= ± c**

Expands to become:

- E	**= - m c(2)**
- E ÷ - m	**= c(2)**
√(- E) ÷ √(- m)	**= √c(2)**
± iE(1/2) ÷ ± im(1/2)	**= ± d / +/- t**

It has been stated: The two sequences of graphics demonstrate the validity of the metaphysical system of the individual acting within God. Einstein's equation intuitively demonstrates the relationship which exists between the abstract and the physical.

How do we move from the intuitive to the demonstratively obvious?

How do we demonstrate:

1. If the opposite of energy (- E) and the opposite of matter (- m) are equated to the speed of light in a vacuum, what could this possibly mean metaphysically?
2. If energy and matter are equated to the speed of light in a vacuum, what could this possibly mean metaphysically?

And finally:

3. Since 'i' is not present in half the graphics: What do 'i' and the lack of 'i' have to do with the metaphysical understanding regarding the whole of reality?

---

28. The real:

The real does not encompass the 'i'. Rather the real encompasses the lack of 'i' which allows the 'real' illusion to deal with the 'i'.

We find the real, from our point of perception, from the point of perception located within the physical, located within the mathematics of:

E = m c(2)

E ÷ m = c(2)

√E ÷ √m = √c(2)

± E(1/2) ÷ ± m(1/2) = ± c

Then if c is velocity and if velocity is the coefficient of constancy k which equals 1:

± E(1/2) ÷ ± m(1/2) = ± k

± E(1/2) = ± m(1/2)

Since we are not interested in the cardinal value of E and m we can ignore the factor of ½ we can concentrate upon the positive and negative metaphysical significance of existence, we can concentrate upon the four scenarios:

+ E = + m

+ E = - m

- E = + m

- E = - m

The four scenarios epitomize the principle of physics dealing with symmetry:

1.	**+ E**	**=**	**+ m**
2.	**+ E**	**=**	**- m**
3.	**- E**	**=**	**+ m**
4.	**- E**	**=**	**- m**

Physics presently has shown energy, matter, and anti-matter to exist. The principle of symmetry would strongly suggest anti-energy would need to exist to complete the symmetry regarding the relationship:

It matter (+ m) plus anti-matter (- m) annihilate each other and produce energy:

+ m + - m = 0 m = E

Then what is the symmetrical equivalent of energy:

+ E + ? = 0 E = ?

The symmetrical equivalent to energy should be anti-energy just as the symmetrical equivalent to matter is anti-matter.

The physical model of reality suggests there are four quadrants to physical reality, four quadrants to 'reality', four quadrants to the 'real', four quadrants to Cartesianism:

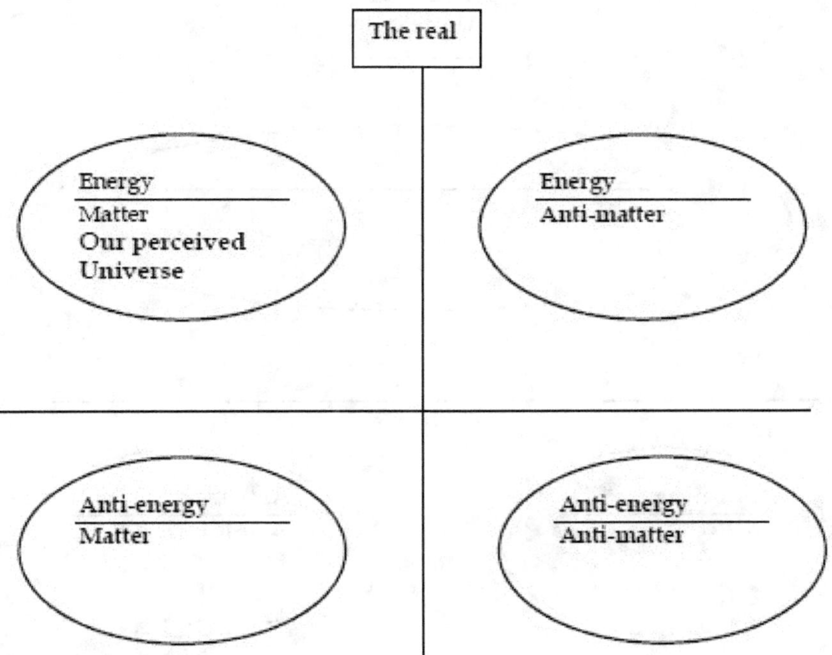

We know matter combined with anti-matter gives us pure energy. As such combining the existences composed of matter and energy with that of existences composed of anti-matter and energy gives us:

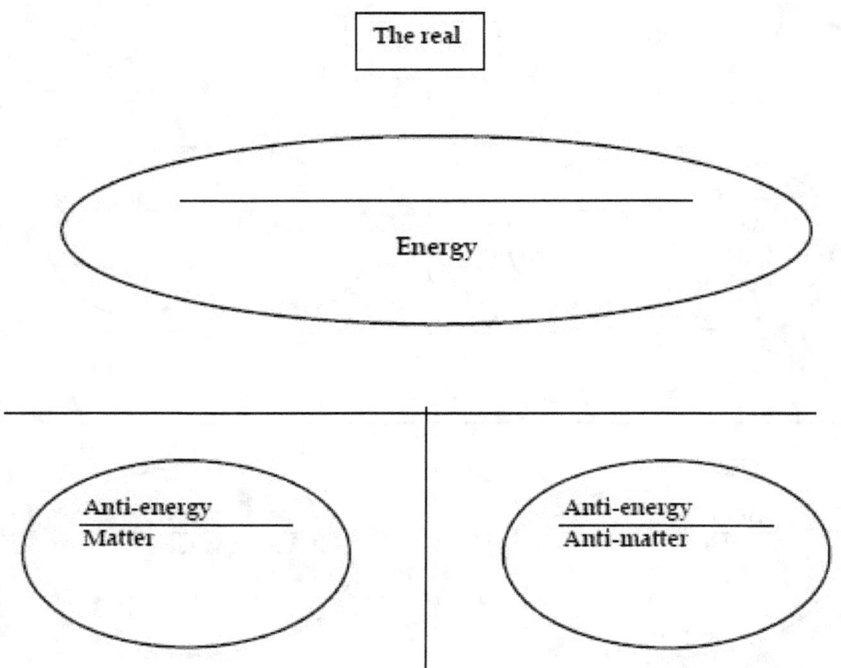

The principle of symmetry would suggest: Combining anti-matter immersed within anti-energy with matter immersed within anti-energy would give us pure anti-energy. As such combining the existences composed of matter and anti-energy with that of existences composed of anti-matter and anti-energy would theoretically give us:

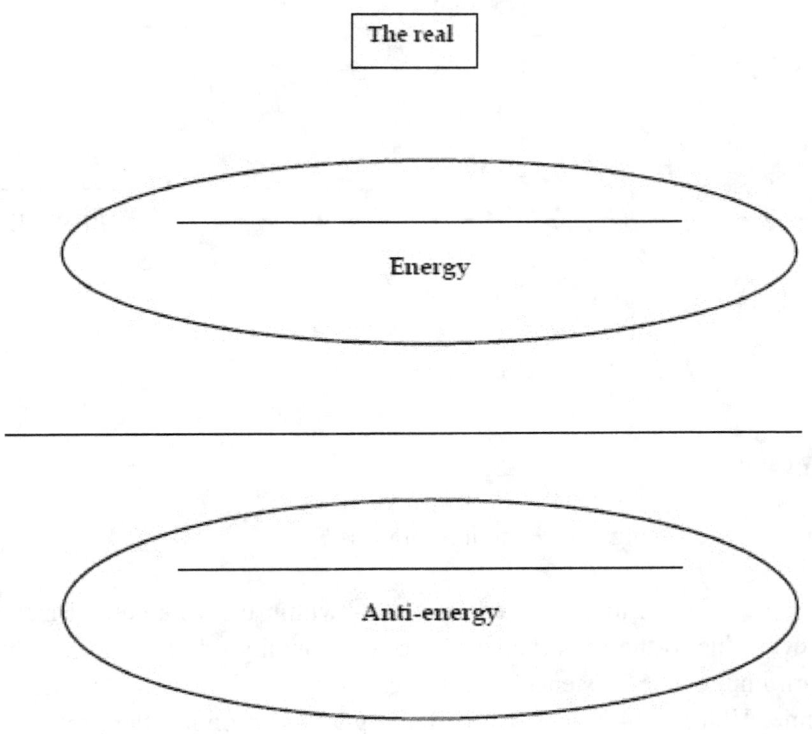

Moving the model one step further: The union of energy and anti-energy should give us nothing:

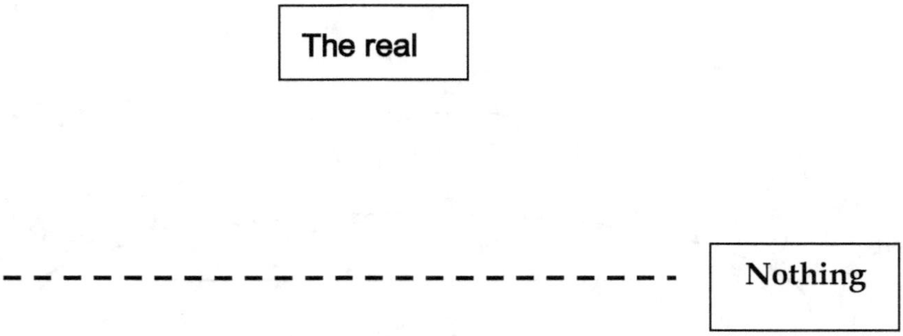

The real

- -

Nothing

How can this be?

How can 'everything' reduce to nothingness?

The topic is much too involved to discuss within the volume of Einstein, however, due to the importance of the functionality of nothingness which in turn implies the existence of nothingness, the topic will capture its own volume: Volume 14: The Error of Heidegger – Resolving the problem of the void of a void. In addition, it may provide some form of validity to such a concept as the dissolving of universes into nothingness, if we at this point add that billions of people have adhered to just such a concept.

The ancient and still functioning religion of Hinduism holds to just such a belief.

Although this volume does not deal with ontological concepts as such, the mention that persistent 'universal/timeless' (earth oriented) religious, mythical, and humanly intuitive ideas are now thought to provide our species with some form of direction and reality based insights regarding the very reality within which we find ourselves to exist.

For now, however, lets focus upon the idea that symmetry plays an important part regarding the equation of reality as expressed by panentheism as opposed to panentheism.

The principle of symmetry would suggest we will find a physical presence identified as anti-energy.

Whether we find the presence of anti-energy as an actual physical presence or we find anti-energy cannot exist within the presence of energy but the mathematical expression of the existence of anti-energy can be demonstrated is not the point.

What is the point is that we should be able to demonstrate the existence of such a 'thing' as anti-energy eventually if the principle of symmetry applies.

But what if the concept of anti-energy does not emerge as a viable concept?

What if the principle of symmetry is proven by science to be an erroneous concept of our universe?

Will such a development simultaneously eliminate the metaphysical model of non-Cartesianism powered by Cartesianism as a viable model of metaphysics?

The answer is: The development of asymmetry as the principle of the universe would not eliminate non-Cartesianism powered by Cartesianism as a viable model of metaphysics. Volumes 5 – 22 of this work would not all become irrelevant just because speculation regarding the more theoretical examination of Volume 16 proved to be sheer speculation with the advent of two thousand years of theoretical physics having been 'proven' wrong.

Moving on, we must then ask: What is this thing we call 'time', which is found within the realm we call the real, the physical, the realm we occupy as physical 'beings'?

What are the characteristics of time and how does time, found within pockets of regions we call universes and individuality, which in turn are located within the purity of abstraction lacking the universal fabric of time, affect the region of timeless abstraction itself?

28. Coherency of time

Coherency of time, one-dimensional time, much as one dimensional light, moves forward as one vertical band and maintains a dimension forward while simultaneously sliding up and down. Time, therefore, is multi-dimensional in nature.

Time appears to move unimpeded as well as unaffected by 'direction'. Therein lies the coherency of the concept we call time/distance

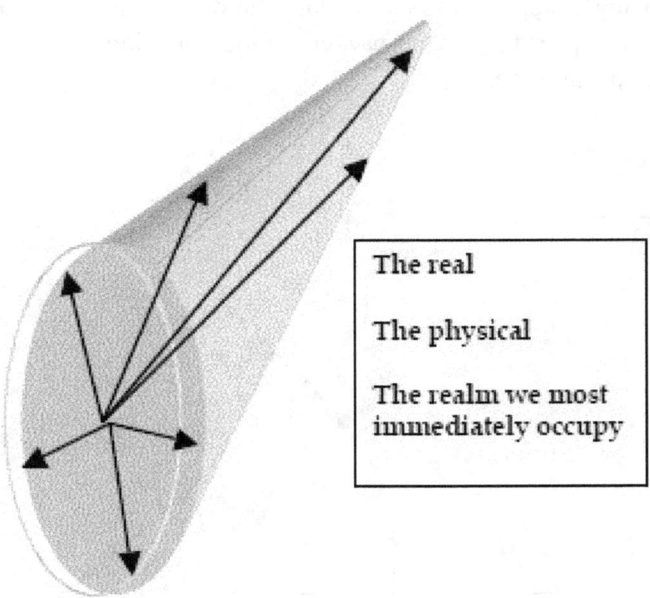

Time is coherent in this aspect since time is found not in packets of time but in a form of continuity.

Time in the concrete, time within the physical, has continuity. As we shall see, time can also lack continuity. Incoherent time, time lacking continuity, is found within the abstract.

Whether the universe is linear or quadratic in nature is not the issue of this volume nor is the issue of this volume the understanding regarding whether time is affected by distance and distance is affected by time.

What is of metaphysical interest within this volume is that time 'appears' to be unaffected by the 'direction' it traverses. What is of interest metaphysically is that we, knowing beings, appear to accumulate knowing in a linear fashion directly associated with the linear movement of time.

We, knowing beings, appear to accumulate 'knowing' as we travel 'through' time. We appear to pick up packets of experiencing wrapped in time. As such we appear to be:

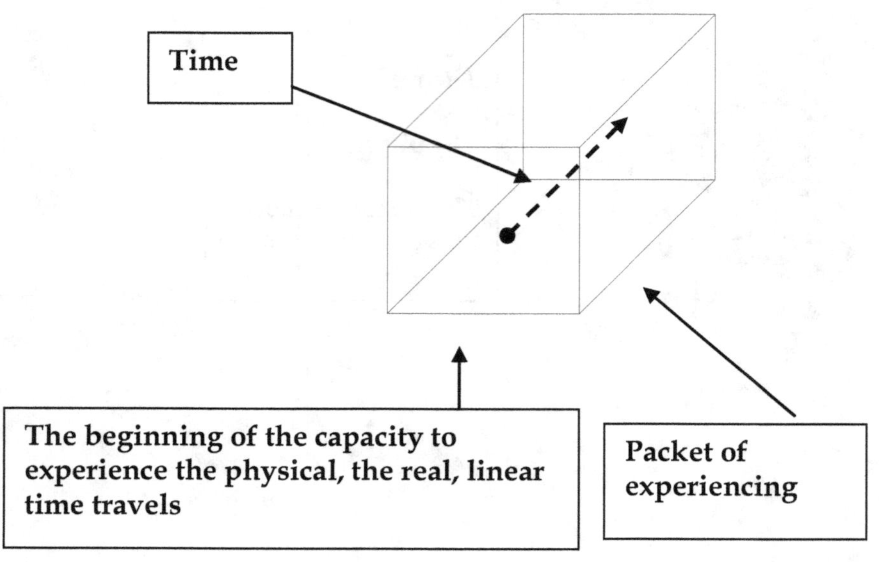

The complete entity of the individual/'knowing' begins with our first being capable of being 'aware' of our experiencing becomes a complete entity upon reaching the end of our being able to experience within this realm we call the physical, reaching the end of our being capable of the individual 'aware' of this realm involving 'what could be', reaching the end of our being personally capable of experiencing through awareness this realm involving linear time

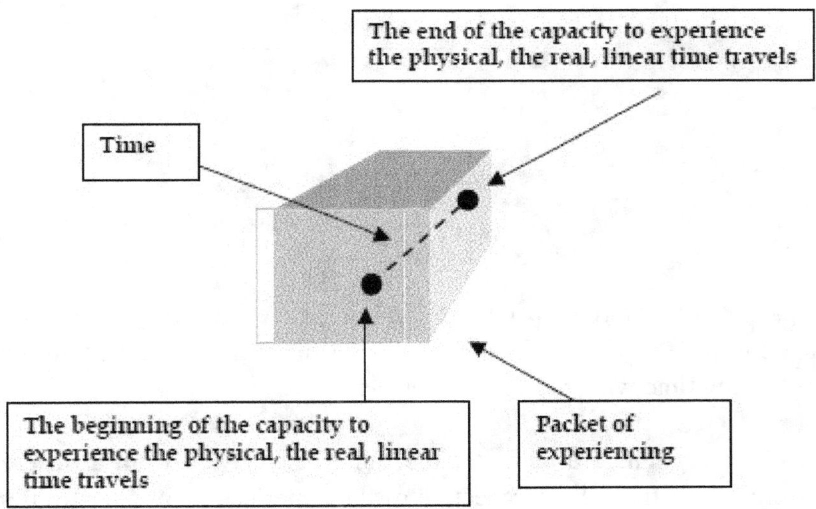

The tunnel of time now becomes:

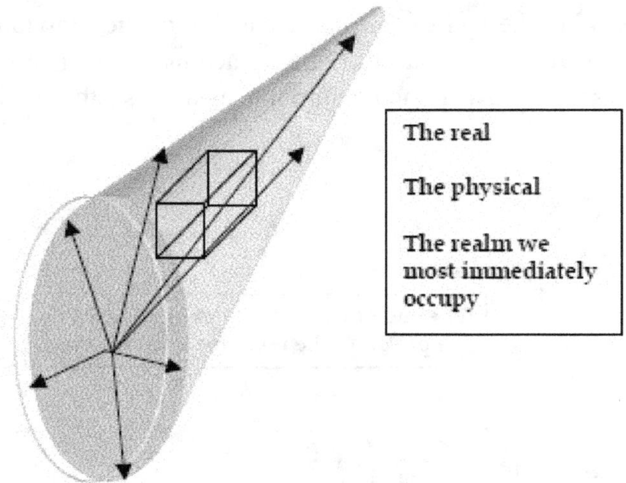

The packet of awareness travels through time gleaning experiences wrapped in time or one could say the packet of awareness travels through time gleaning time wrapped in experiences.

The two perceptions are basically the same just viewed from a different perspective, one from the perspective of the experience the other from the perspective of time.

The question then becomes what happens to these packets of experience wrapped in time/time wrapped in experience?

If the physical is all there is then the answer becomes: What happens to the packets of experience wrapped in time is exactly the same as what happens to the physical itself.

If the physical ‘dissolves’, ‘ends’, so to end the packets of experience.

If the physical is all there is then the packet of experience, the knowing being, is a closed system.

The realm of the physical is closed.

The physical, the universe is closed by the limits of time itself, is closed by the very apparent infinity of time for as surely as time appears infinite it becomes encaged within the limits of the concept of time itself.

28a. Variability of time

As convenient as it may be to suggest time is coherent, consistent, we believe time is variable in nature. Einstein's Theory of Relativity suggests time 'changes' with the rate at which one moves through distance.

Does the variability of time affect the concept of packets of experiencing, the unit of knowing the individual 'accumulating' experiencing as it travels through time? Einstein's theory would suggest such is not the case. Rather it is the experiencing, which is experienced. It is the experiencing as it relates to time that awareness itself experiences.

The result: The packet is the packet, the experiences are the experiences, the concept of 'filling' the packet remains the same regardless of the 'speed' through which the unit of knowing traverses through the physical.

Time becomes encapsulated within the packet and experience becomes just that experience. On the other hand, time does not become 'just' time but rather time becomes the packet, the packaging of the abstractual knowing of the experience.

29. The 'real' illusion

The 'real' illusion does not but does encompass the 'i'.

The 'real' illusion symbolically encompasses the 'i' through the mathematical hieroglyphics which allows the 'real' to deal with the 'real' illusion symbolically through the use of 'i'.

We find the 'real' illusion, from our point of perception, from the point of perception located within the physical, from the point of perception located within the real, located within the mathematics of 'i', becomes symbolized by the mathematical statements:

- E = - m c(2)

- E ÷ - m = c(2)

√(- E) ÷ √(- m) = √c(2)

± iE(1/2) ÷ ± im(1/2) = ± c

Then if c is velocity and if velocity is the coefficient of constancy k which equals 1:

± iE(1/2) ÷ ± im(1/2) = ± k

± iE(1/2) = ± im(1/2)

Since we are not interested in the cardinal value of E and m we can ignore the factor of 1/2 and we can concentrate upon the positive and negative metaphysical significance regarding the existence of four scenarios:

+ iE = + im

+ iE = - im

- iE = + im

- iE = - im

- E = - m c(2)

- E ÷ - m = $\frac{d(2)}{t(2)}$

√(- E) ÷ √(- m) = $\frac{\sqrt{d(2)}}{\sqrt{t(2)}}$

± iE(1/2) ÷ ± im(1/2) = ± d/t

Then if c is velocity and if velocity is the coefficient of constancy k which equals 1:

± iE(1/2) ÷ ± im(1/2) = ± 1

Existing as we do within the physical, we can, through the previous progression of thought suggested in the section: The real, somewhat understand the mathematical concepts:

E = m c(2)
E ÷ m = c(2)
√E ÷ √m = √c(2)
± E(1/2) ÷ ± m(1/2) = ± c

Then if c is velocity and if velocity is the coefficient of constancy k which equals 1:

± E(1/2) ÷ ± m(1/2) = ± k
± E(1/2) = ± m(1/2)

Since we are not interested in the cardinal value of E and m we can ignore the factor of 1/2 and we can concentrate upon the positive and negative metaphysical significance regarding the existence of four scenarios:

+ E = + m
+ E = - m
- E = + m
- E = - m

Reduces to:

E	**= m c(2)**
E ÷ m	**= c(2)**
√E ÷ √m	**= √c(2)**
± E(1/2) ÷ ± m(1/2)	**= ± c**
Then if c is velocity and if velocity is the coefficient of constancy k which equals 1:	
± E(1/2) ÷ ± m(1/2)	**= ± 1**

It would appear, through the thought process outlined in both the section, 'The real' and 'The real illusion', that we have two scenarios. This perception, however, is incorrect. We actually have four scenarios:

We have the two scenarios outlined:

E	= m c(2)
E ÷ m	= c(2)
√E ÷ √m	= √c(2)
± E(1/2) ÷ ± m(1/2)	= ± c
Then if c is velocity and if velocity is the coefficient of constancy k which equals 1:	

- E	= - m c(2)
- E ÷ - m	= c(2)
√(- E) ÷ √(- m)	= √ c(2)
± iE(1/2) ÷ ± im(1/2)	= ± c
Then if c is velocity and if velocity is the coefficient of	

And we have:

E	$= - m\ c(2)$
$E \div - m$	$= c(2)$
$\sqrt{E} \div \sqrt{(- m)}$	$= \sqrt{c(2)}$
$\pm E(1/2) \div \pm im(1/2)$	$= \pm c$

Then if c is velocity and if velocity is the coefficient of constancy k which equals 1:

$- E$	$= + m\ c(2)$
$- E \div + m$	$= c(2)$
$\sqrt{(- E)} \div \sqrt{(+ m)}$	$= \sqrt{c(2)}$
$\pm iE(1/2) \div \pm m(1/2)$	$= \pm c$

Then if c is velocity and if velocity is the coefficient of constancy k which equals 1:

Symmetry prevails but what is the symmetry suggested? We leave the physics to the physicists but we will tackle the metaphysics of the symmetry suggested since this work, The War and Peace of a New Metaphysical Perception, deals with metaphysics.

The concept of the 'real' and the 'real' illusion enters the picture as the means of best describing the symmetry implied by the mathematical symbolization suggested within the equations.

The real – the lack of 'i', and the 'real' illusion – the presence of 'i' appear an equal number of times within the four scenario final equations.

The potential explanation lies in Zeno's description of the existence of the abstract and the physical. In short we have the physical and the abstract.

We come full circle to Volume 5: The Error of Zeno – Resolving the problem of The Abstract.

It appears a 'mirror' exists separating the real from the 'real illusion, separating the physical from the abstract, allowing the paradox of motion expressed by Zeno to be realized as two equally dynamic arenas of existence for elements of knowing producing two equally significant purposes for elements/multiplicity in terms of how they affect the whole/singularity.

The ‘mirror’ is revealed through the examination of Einstein’s equations regarding the relationship of matter and energy. The ‘i’ and the lack of ‘i’ evolve as the expressions of the two locations of existence.

We have observed the emergence of the rationale supporting two regions one within the other, the ‘contained’/Cartesian/the closed system allowing the ‘container/the non-Cartesian/the open system to ‘grow’ versus the Aristotelian perception of the ‘container’ existing in a state of decay or existing in an eternal state of permanent equilibrium.

We have observed the emergence of the rationale behind the concept of a non-Cartesian system of abstraction ‘containing’ a Cartesian system of physicality wherein the Cartesian system, ‘powers’, provides the mechanism by which the non-Cartesian whole thrives, flourishes, blossoms, lives, avoids the very nihilistic purposelessness of Nietzsche’s ‘eternal recurrence’.

The separation:

Aristotle's Cartesianism Cartesianism	Kant/Hegel's non-
The Physical	The Abstract
Cartesianism	Non-Cartesianism
Free Will	Determinism
Centricism	Non-Centricism
Finite Time In Infinite Space	Infinite Time In Finite Packets
Finite Distance In Infinite Time	Infinite Distance In Finite Packets
What 'could be'	What 'was', What 'is', What 'will
Closed System	Open System
Coherent Time	Incoherent Time
Active Observation	Passive Observation

Now becomes a hybrid of the two:

Aristotle's Cartesianism Cartesianism	Kant/Hegel's non-
The Physical	The Abstract
Cartesianism	Non-Cartesianism
Free Will	Determinism
Centricism	Non-Centricism
Finite Time In Infinite Space	Infinite Time In Finite Packets
Finite Distance In Infinite Time	Infinite Distance In Finite Packets
What 'could be'	What 'was', What 'is', What 'will
Closed System	Open System
Coherent Time	Incoherent Time
Active Observation	Passive Observation

In the form of two interdependent systems fused together as an energetic viable entity of wholeness. The system more clearly becomes:

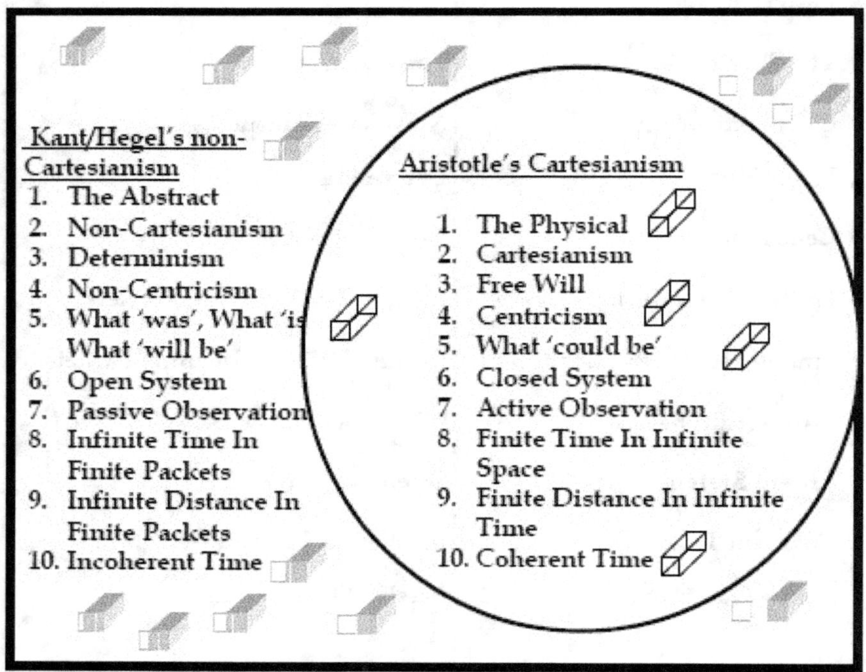

Kant/Hegel's non-Cartesianism
1. The Abstract
2. Non-Cartesianism
3. Determinism
4. Non-Centricism
5. What 'was', What 'is' What 'will be'
6. Open System
7. Passive Observation
8. Infinite Time In Finite Packets
9. Infinite Distance In Finite Packets
10. Incoherent Time
Aristotle's Cartesianism
1. The Physical
2. Cartesianism
3. Free Will
4. Centricism
5. What 'could be'
6. Closed System
7. Active Observation
8. Finite Time In Infinite Space
9. Finite Distance In Infinite Time
10. Coherent Time

The questions become: What is meant by items numbered 8, 9. and 10 found within the diagram? What do: Infinite Time In Finite Packets, Infinite Distance In Finite Packets, Finite Time In Infinite Space, Finite Distance In Infinite Time, Coherent Time, and Incoherent Time, have to do with the metaphysical understanding regarding the Newtonian 'i' and the Einsteinian 'i'?

Metaphysically, positive and negative energy and positive and negative matter interpret into the potential to be energy and anti-energy and matter and anti-matter.

Other than anti-energy, physics deals with these concepts daily. Positive distance and positive time are concepts with which we all deal on a daily basis.

What, however, of the concept of a negative distance and a negative time? Negativity of time and negative distance become a mathematical means of expressing inverse relationships.

From the section of this volume entitled: The square root of Einstein's equation: 'i' , we saw the initiation regarding the concept of a 'negative' time and a 'negative' distance.

The schematic gives us eight options.

From the first set of equations we obtain:

1. $c = E/m$
2. $c = -E/m$
3. $c = E/-m$
4. $c = -E/-m$

From the second set of equations we obtain:

5. $-c = E/m$
6. $-c = -E/m$
7. $-c = E/-m$
8. $-c = -E/-m$

Having examined such concepts in detail, we can now begin to examine the concept referred to as the 'incoherency of time'.

29a. Incoherency of time - reiterated

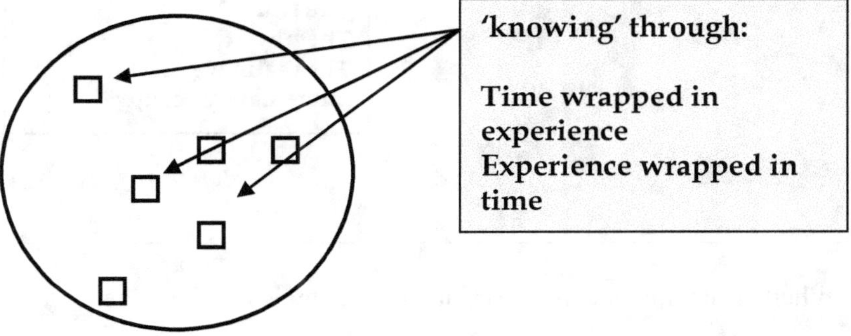

Three dimensionally we obtain:

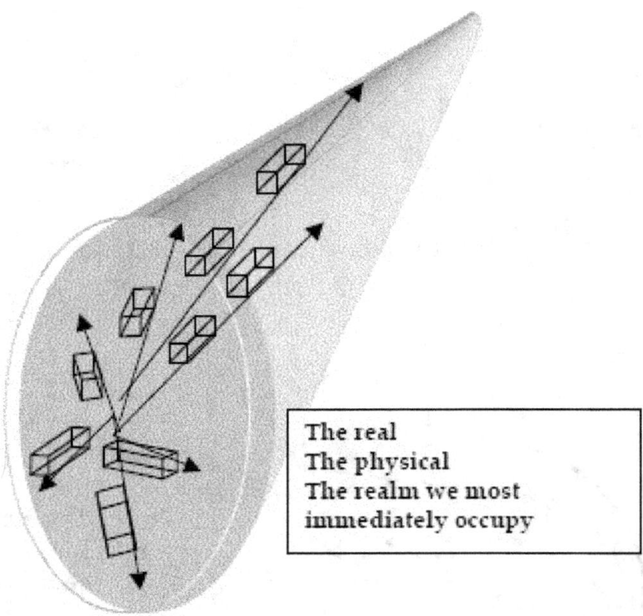

Whether the universe is linear, quadratic, etc, is not the issue.

It may be possible time is affected by distance and distance by time but that also is not the point of this volume.

What is of interest metaphysically is that time 'appears' to be unaffected by the 'direction' it traverses.

What is of interest metaphysically is that we, knowing beings, appear to accumulate knowing in a linear fashion directly associated with the linear movement of time.

We, knowing beings, appear to accumulate 'knowing' as we travel 'through' time.

Again we understand: We appear to pick up packets of experiencing wrapped in time. As such we appear to be:

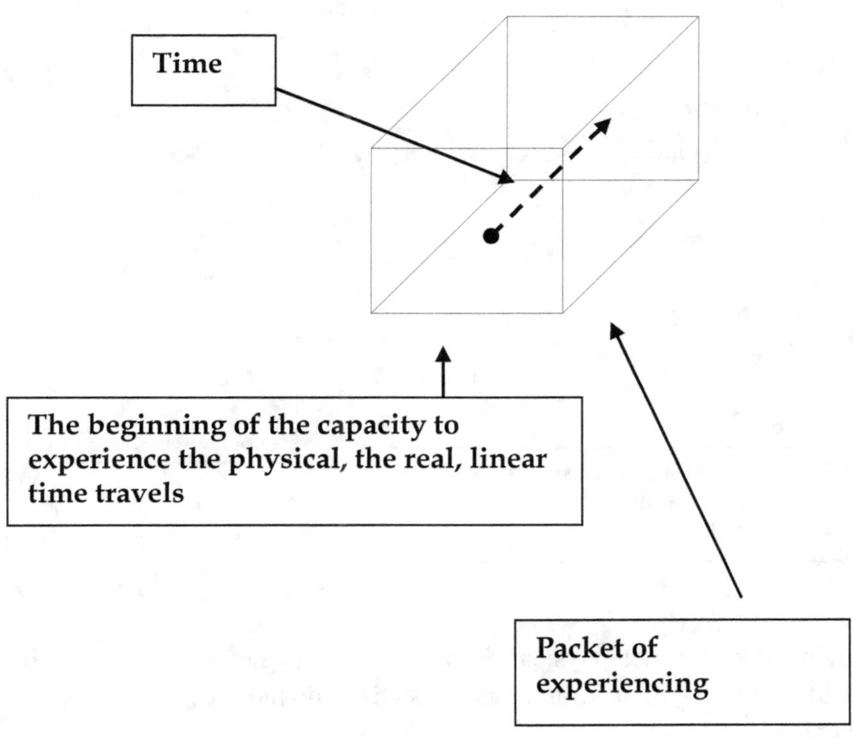

Beginning with our first being capable of being 'aware' of our experiencing to the end of our being able to experience within this realm we call the physical, this realm of 'what could be', this realm of linear time becomes complete and thus the graphic of the completed unit becomes:

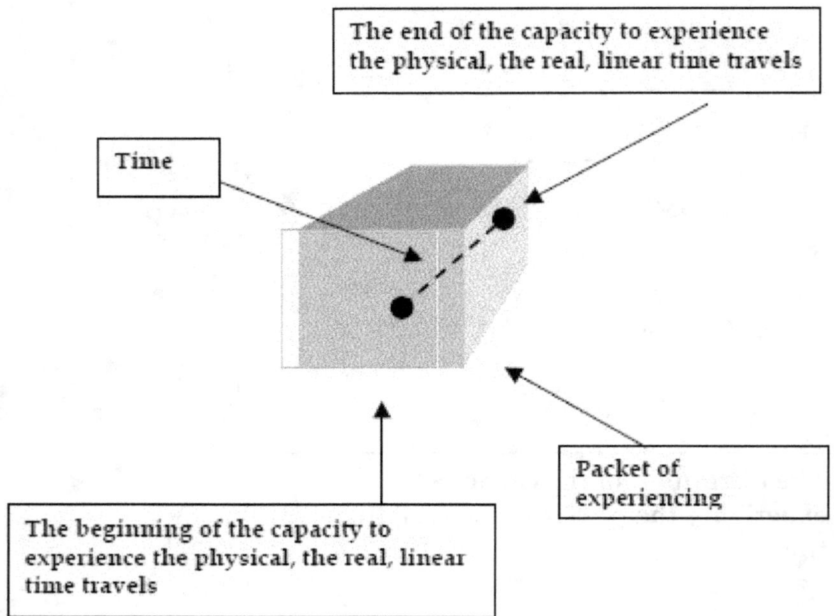

Having laid the metaphysical foundation, we begin to understand there should logically be a symmetrical 'location for the incoherency of time. There should be a location of 'inconsistency of time.

Such a location would not appear to be 'within' the coherency of time but rather be 'an independent' location of incoherency of time containing the coherency of time.

Such a 'location' would in actuality be a location unlimited by the parameters of space and time.

But how can one have a location void space and time? The very 'lack' of space and time implies no 'location'.

Perhaps, but such suggestions spring from an inability to perceive of 'a location' void the universal fabric of space and time due to the fact that such a concept is 'alien' to our thought process.

If we were to diagram such a 'location', we might do so as:

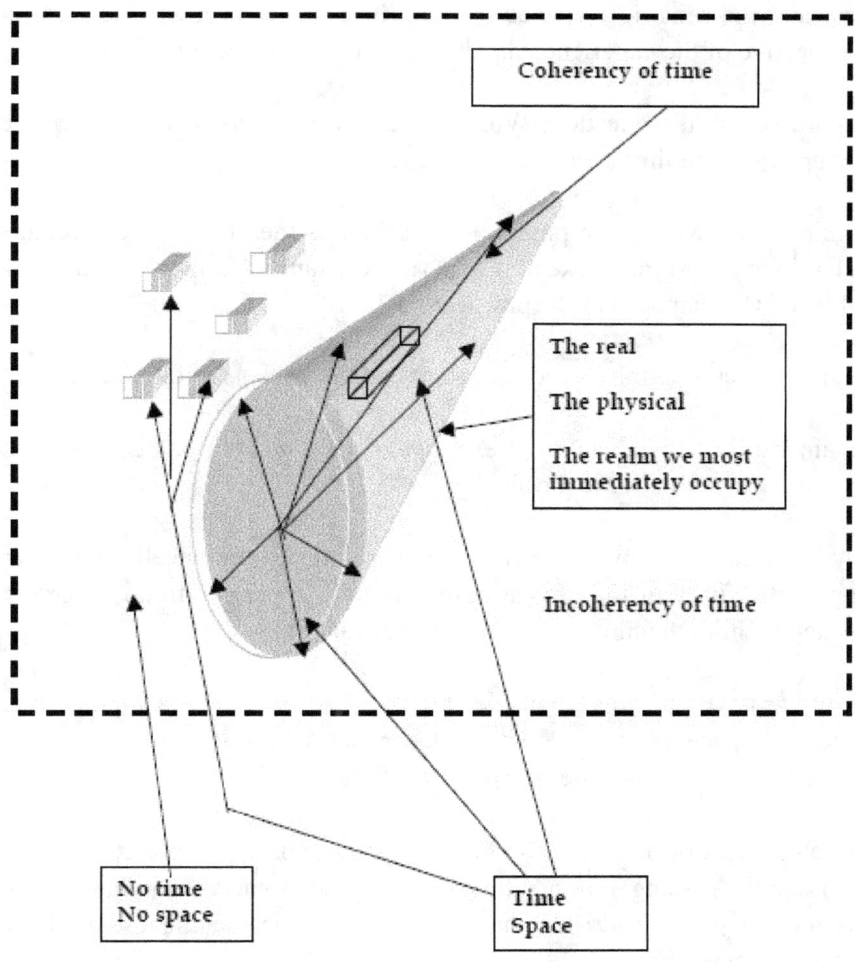

The packet of awareness travels through time gleaning experiences wrapped in time or one could say the packet of awareness travels through time gleaning time wrapped in experiences. The two perceptions are basically the same just viewed from a different perspective, one from the perspective of the experience the other from the perspective of time.

Again we ask the question: What happens to these packets of experience wrapped in time/time wrapped in experience?

Again we answer: If the physical is all there is then the answer becomes: What happens to the packets of experience wrapped in time is exactly the same as what happens to the physical itself.

If the physical 'dissolves', 'ends', so to end the packets of experience.

Again we state: The packet of experience, the knowing being, is a closed system. The realm of the physical is closed.

The physical, the universe is closed by the limits of time itself, by the very apparent infinity of time for as surely as time appears infinite it becomes encaged within the limits of the concept of time itself.

Now we have an 'area' 'outside' the physical. We have a region where time and space are found 'within' the packets of knowing but time and space do not exist 'outside' the packets of knowing.

For awareness of one packet to travel to another packet of awareness is not an issue of passing through time nor is it an issue of passing through distance. It just is and as such 'travel' becomes instantaneous in the true sense of instantaneous.

The linear sequencing of time is found only 'within' the packets and not between packets. As such the concept of linear movement from one time frame to another does not exist external to the packets but only exists as an internal function.

The result is a form of non-sequentiality of time taking place as a universal medium. Eternal time takes on the characteristic of incoherency.

All is capable of being experienced within the realm of the incoherency of time with the exception of 'what is not, with the exception of 'what could be' for 'what could be' is not yet.

Does an entity of multiplicity experiencing another's experience 'affect' either entity?

Within such a model the answer is: 'Experiencing occurring after the entity has been fully constructed, completed as an entity, does not 'change' the 'complete' entity of knowing.

The only way to 'change' is to experience 'what one constructs through the conversion of 'what could be' to being 'what is'.

Thus the region of 'what could be' becomes the engine, the 'power' source of 'what is'.

In essence we have a non-Cartesian system powered by the Cartesian.

We have the individual acting within God as well as the individual acting within God.

Again it will be stated that it is the first relationship, the individual acting within God, that is the issue of metaphysics and thus the issue of this work, The War and Peace of a New Metaphysical Perception.

It is not the relationship of the second, the individual acting within God, that is the issue with which Metaphysics is concerned.

29b. Constancy of time

As convenient as it may be to suggest time is incoherent, inconsistent, within the realm of the purity of abstraction, it is possible for time to be absent as a 'universal' fabric of the abstract and thus inconsistent in its lack of consistency when viewed as being absent.

In such a region, Einstein's Theory of Relativity would suggest time does not 'change' with the rate at which one moves through distance.

Does the lack of time, does travel 'through' timelessness affect the concept regarding packets of experiencing the unit of 'knowing being'.

To travel 'through' timelessness would appear to have no affect upon the unit of knowing since without time the unit of knowing composed of experiences 'created' in time would be unaffected by a timeless realm.

What then of the travels one unit of knowing may 'accumulate' as it travels through time found within the packets of knowing found to exist within the realm of timelessness?

Would such an 'experience' affect the original unit of knowing doing the transposition upon the new unit of knowing?

No it does not.

The unit of knowing doing the transposing may experience the event of new experiencing and fully sense the new experience but the event would not 'change' the unit of knowing for it is what it is and cannot 'change' 'what it is', 'what it was', nor 'what it will be' for there is no universal fabric of time present through which the development of 'what could be' can develop.

The unit of knowing remains as it is.

The units of knowing 'beings' no longer travel a region of time to develop their own unique form of knowing and thus it appears they could not become new unique entities from other knowing.

The result: The packet is the packet, the experiences are the experiences, the concept of 'filling' the packet remains the same regardless of new understanding these unique packets of knowing glean as they complete their new travels within the region of time found to exist as isolated packets within the realm of timelessness.

Units of knowing can traverse the lack of no distance in no time since no time exists between units of knowing and no distance separates one packet of knowing from another.

Having traversed non-existent dimensions of time and distance in no time at all, individuality, units of unique knowing, packets of multiplicity, can then proceed to experience other forms of experience having been established by other forms of unique perceptions which in turn have been developed through their unique travels through time experienced by virgin consciousness.

The two, the constancy of time and the incoherency of time, found within the 'real illusion' become the mirror images the two, the variability of time and the coherency of time, found within the 'real'.

---

30. The 'Taser':

Taser: time amplification by stimulated emissions of radiation – as opposed to laser

The radiation: packets of time much as the packets of light – photons of light but in this case packets of time

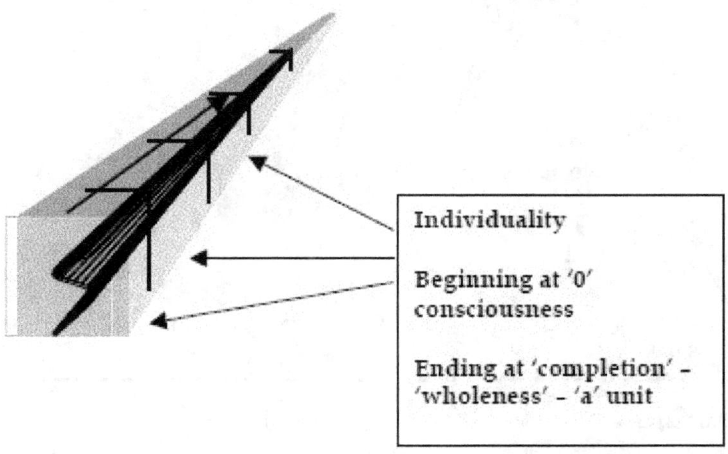

Individuality begins to appear to lose its very uniqueness of individuality the further one removes themselves from individuality but in essence individuality is still what it is – individuality 'within' which time is immersed, wrapped.

Pull such individuality apart and let it float randomly and one obtains an incoherency of time, an incoherency due to apparent chaotic movement, Brownian movement of packets of time within which coherency of time exists as the very structure of the individual's structure itself.

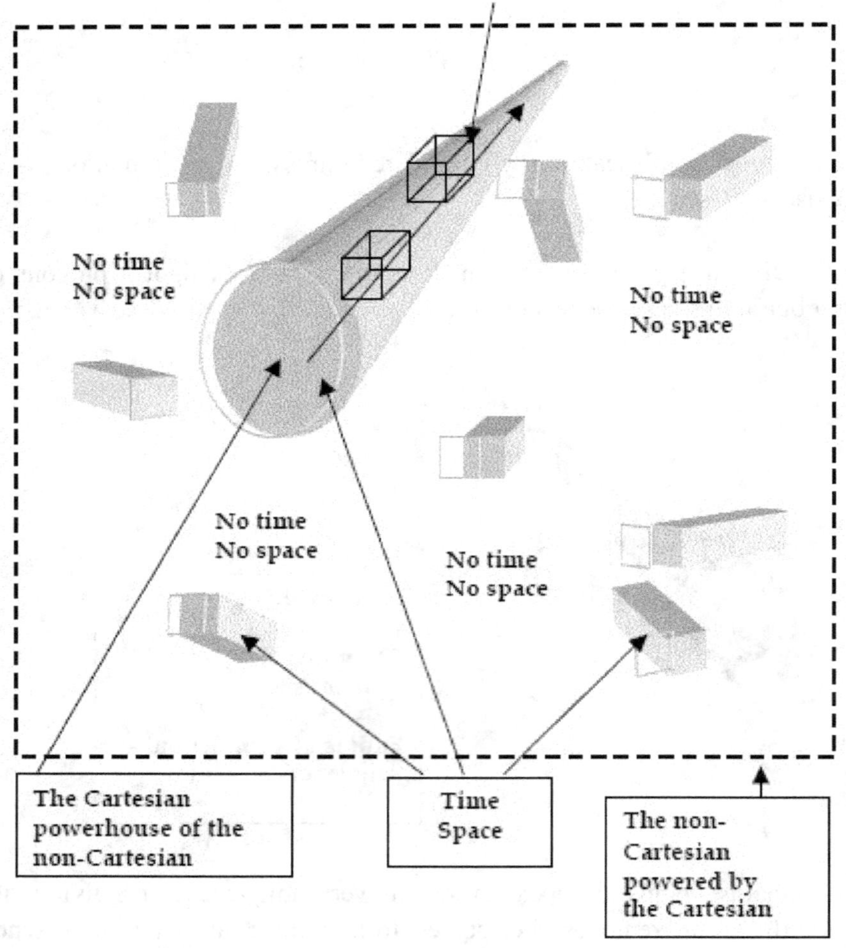

The multiplicity of knowing unique personalities of experiencing are capable of movement within the non-Cartesian much as packets of light have a pattern of random, incoherent movement within the physical.

Take the packets of light and cause them to move from an incoherent movement pattern to a coherent movement pattern and one obtains a 'laser' light beam, light amplification by stimulated emissions of radiation.

What does this have to do with time and the concept of two 'locations' of existence?

The analogy of a laser can be used to best describe what time represents to the purity of abstraction. Time takes the place of light in the analogy. Time replaces light and thus we obtain the concept of 'taser': time amplification by stimulated emissions of radiation.

Just as the laser becomes a unique and powerful tool for our purposes within the physical, within the real, within our universe, so time becomes a unique and powerful tool for the purposes of the summation of knowing, for the purposes of the whole, for the purposes of God as opposed to the individual.

The 'taser' becomes a means of developing virgin consciousness into a functional entity with a purpose to the whole.

The 'taser' becomes a means by which the entities of knowing in the form of multiplicity of individuality mold themselves into unique complete packets of experiencing wrapping time within its folds. The objective: to keep the summation of knowing,

God from existing in a constant state of permanent equilibrium stagnating in the quagmire of 'eternal recurrence'.

In short, the 'taser' becomes the means by which the whole, God, the summation of knowing, becomes even more so without the coherency of time's linear movement being a limiting factor to the whole itself other than in the sense of the whole being 'freed' to become more than it is, was, and will be.

Now the whole can become what it might be, could be, but not necessarily 'will be'.

The possible perception that the process of time amplification is a simplistic concept is misleading.

There is no doubt time is a complex subject of which we know relatively little. This volume does not imply an in-depth understanding regarding the physics of time.

Metaphysics is not intended to replace physics.

Metaphysics is intended to think 'ahead' of physics and lay groundwork 'towards' which thinkers of physics, the theoretical physicists, can find intriguing ideas and concepts for their thinking process.

Metaphysics has been reticent in its function to provide the fodder of thought for theoretical physicists.

The blame for the relatively stagnant state of mind regarding metaphysical thinkers lies not with science but rather the blame lies with philosophy and philosopher.

The state of existence within which 'politically correct' institutions and philosophical thinkers have been compressing metaphysics itself into an ever decreasing sphere of influence.

It is the reversal of this process of compression, which is the goal of this work, The War and Peace of a New Metaphysical Perception.

But what of time and our future examination of time?

The complexity of time explodes with the knowledge we gain of the abstractual characteristics of time itself.

For example:

Time may be found within time itself.

Tunnels of time may be found within tunnels of time

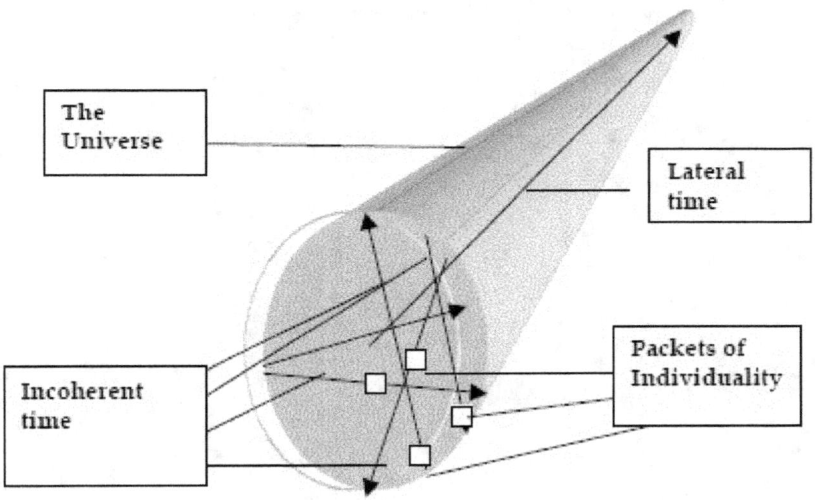

Packets of individuality if given 'enough' space, could move in different time directions relative to each other in perspective of viewing one galaxy from another

Our own propulsion systems thrusting us at phenomenal speeds of one individual relative to another may create what are perceived to be 'Doppler effects of time' as opposed to 'Doppler effects of sound'. (In effect these 'perceived' variations of time may be but the illusion of 'Doppler effects of time' created by the packets of individuality passing each other in their process of ' being', in their journey of 'life', newly experiencing, filling themselves, *being.)*

The universe, in *being* itself, moves in its own time direction while encompassing individual packets of time moving within it. As such, universe's may journey to allow creation to 'become' so that creation of newness itself moves from potentiality to the individual.

---

31. What does it mean:

It means the significance of the 'constant' variable

E	=	mc(2)
E/m	=	c(2)
E/m	=	kd/t

can be metaphysically demonstrated by the new metaphysical system, and only by the new metaphysical system of the individual acting within God.

It means the significance regarding the relationship of the positive and the negative of both energy and matter …

E	**= m c(2)**
E ÷ m	**= c(2)**
√E ÷ √m	**= √c(2)**
± E(1/2) ÷ ± m(1/2)	**= ± c**

Then if c is velocity and if velocity is the coefficient of constancy k which equals 1:

± E(1/2) ÷ ± m(1/2)	**= ± 1**

… can be metaphysically demonstrated by the new metaphysical system, and only by the new metaphysical system of the individual acting within God.

It means the significance regarding the relationship of the positive 'i' and the negative 'i'

- E	**= - m c(2)**
- E ÷ - m	**= c(2)**
√(- E) ÷ √(- m)	**= √ c(2)**
± iE(1/2) ÷ ± im(1/2)	**= ± c**
Then if c is velocity and if velocity is the coefficient of constancy k which equals 1:	
± iE(1/2) ÷ ± im(1/2)	**= ± 1**

E	**= - m c(2)**	**- E**	**= + m c(2)**
E ÷ - m	**= c(2)**	**- E ÷ + m**	**= c(2)**
√E ÷ √(- m)	**= √c(2)**	**√(- E) ÷ √(+ m)**	**= √ c(2)**
± E(1/2) ÷ ± im(1/2)	**= ± c**	**± iE(1/2) ÷ ± m(1/2)**	**= ± c**
Then if c is velocity and if velocity is the coefficient of constancy k which equals 1:		**Then if c is velocity and if velocity is the coefficient of constancy k which equals 1:**	
± E(1/2) ÷ ± im(1/2)	**= ± 1**	**± iE(1/2) ÷ ± m(1/2)**	**= ± 1**

can be metaphysically demonstrated by the new metaphysical system, and only by the new metaphysical system of the individual acting within God.

It means the significance of the inverse of time and the inverse of distance

$$\frac{1}{kt} = \frac{d}{d}$$

can be metaphysically demonstrated by the new metaphysical system, and only by the new metaphysical system of the individual acting within God.

It means the significance of the seamlessness of distance and the multiplicity of distance

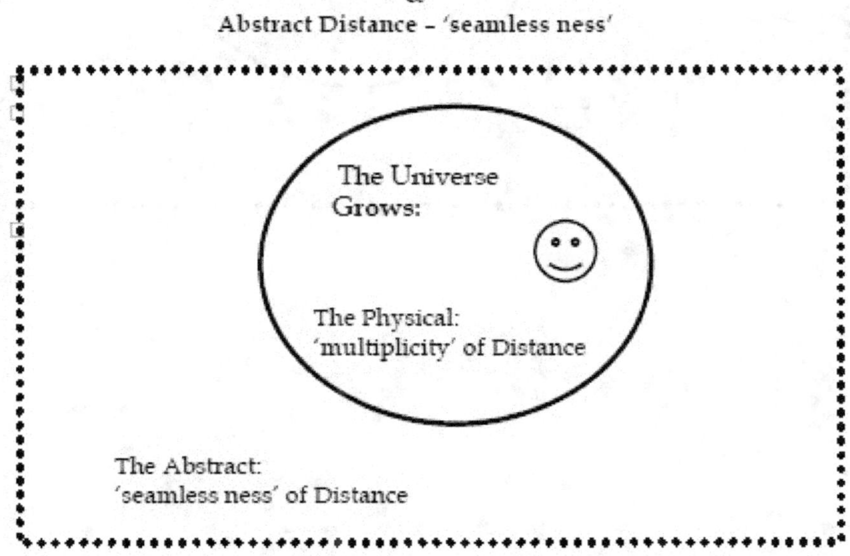

can be metaphysically demonstrated by the new metaphysical system, and only by the new metaphysical system of the individual acting within God.

It means the significance of the multiplicity of time and the seamlessness of time

Abstract Time - 'multiplicity'
&
Physical Time - 'seamless ness'

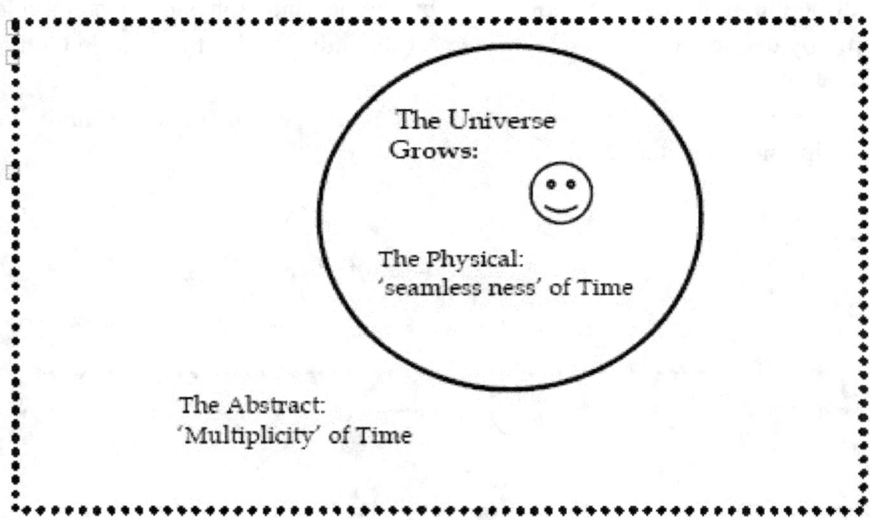

by the new metaphysical system, and only by the new metaphysical system of the individual acting within God.

It means the relationship of the two, time and distance

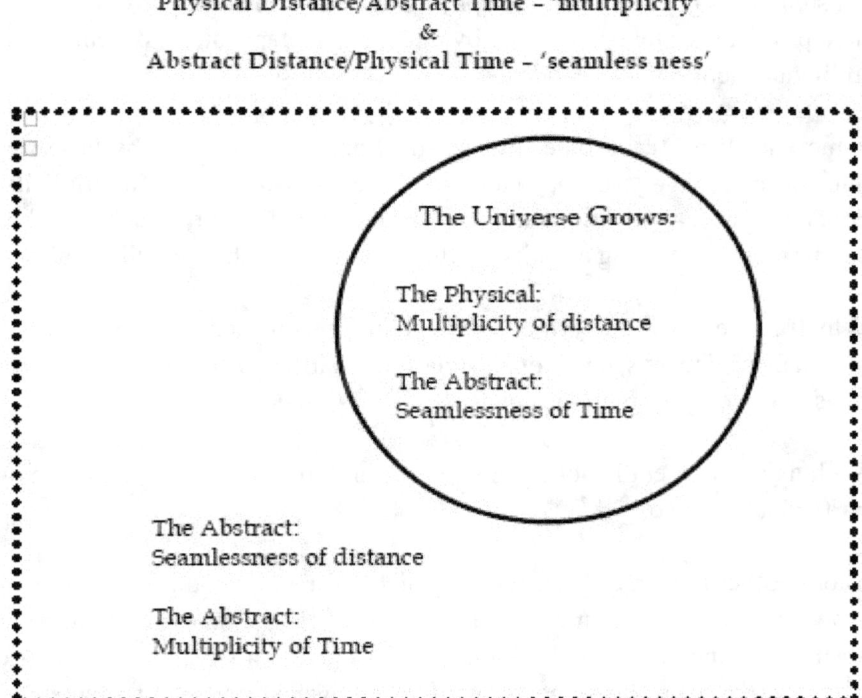

can be metaphysically demonstrated by the new metaphysical system, and only by the new metaphysical system of the individual acting within God.

It means the relationship existing between 'the constancy of time and the incoherency of time' and 'the variability of time and the coherency of time' can be metaphysically demonstrated by the new metaphysical system and only by the new metaphysical system of the individual acting within God.

It means the square of the distance divided by the square of time as it relates to energy divided by mass can be metaphysically demonstrated by the new metaphysical system and only by the new metaphysical system of the individual acting within God.

It means the relationship of the element 'i', the Newtonian 'i', and the Einsteinian 'i' can be metaphysically demonstrated by the new metaphysical system and only by the new metaphysical system of the individual acting within God.

Time and distance become inverse functions of each other because time and distance have two characteristics, 'seamless ness' and 'multiplicity', concrete/physical functionality and abstract functionality, each of which can be 'found' existing as pairs 'within' the 'real' or the 'real illusion'.

Which is the 'real' and which is the 'real illusion' depends upon 'where' one stands, depends upon one's relative position to the abstract and the physical as one refers to the concepts.

Such a development is not unique to time and distance which permeate our particular universe.

Should other universes exist within which abstractions other than time and space are the universal medium, the universal fabric, then unique forms of abstractual knowing found within these unique forms of universal fabrics would experience the same form of metaphysical process as that which occurs to knowing found within the universal fabric of time and space which permeate our particular universe.

The significance of such a perception is that a 'universal philosophy' is a rational concept applying to 'universes' rather than simply our universe. Thus the concept of 'universal' becomes truly universal for it is not applicable to only one universe but to all universes.

The new perception 2000 ad: $v = d/t$ becomes $vt = d$ or $kt = d$, where 'v' equals the constant 'k' and 'k' is not a constant of mathematics but a constant of perceptual existence 'within' a 'real illusion'

But enough has been said regarding the extreme complexity regarding what mathematics and metaphysics have in common as inadvertently implied by Einstein and his concepts of relativity.

Although we are done with Einstein, we are not done with the complexities regarding the relationships which are exposed by Russell and his paradox of members and nonmembers.

In the initial sections of this volume it was stated: Darwinian biologists would say the egg came first.

Creationist biologists would say the chicken came first.

Again and again the question becomes: Which came first the chicken or the egg, the creator or the created?

Metaphysically it might appear we are no further along than we ever were.

The appearance of the lack of metaphysical progress, however, is not an accurate evaluation regarding our metaphysical progress as a species.

We have come a long way from the metaphysical Aristotelian Cartesian understanding of our reality.

We have come a long way from the metaphysical Kant/Hegelian non-Cartesian understanding or our reality.

The resolution to the puzzling paradox regarding Zeno's space/time and space/distance is beginning to evolve nicely.

The new metaphysical perception of the individual acting within God, non-Cartesianism powered by Cartesianism, the Cartesian being a functional part of the whole, a functional aspect of non-Cartesianism, is beginning to provide us with some interesting insights regarding our potentially to understand both our reality and the frontiers of our expanding reality.

The new metaphysical perceptions introduced by this work, The War and Peace of a New Metaphysical Perception, recognize the existence of abstraction existing both timelessly and void the restraints the dimensions time and space/distance impose upon the concept of existence itself.

As such abstractual concepts exist and abstractual concepts grow in number through the process of abstraction expanding itself through the process of traveling isolated from itself by the medium of nothingness epitomized through the temporary form of the physical without which the abstract would stagnate eternally in the psychotic breeding grounds of nihilistic permanent equilibrium, eternal recurrence.

The introduction of the concept, 'nothingness', appears to consistently 'pop-up' inconsistently and inappropriately. The concept of nothingness, however, has been a part of every volume to this point.

It will soon be time to focus upon this most elusive aspect of the new metaphysical model: the individual acting within God. It will soon be time to focus upon the concept of nothingness itself.

First, however, we will examine Russell's paradox regarding members and non-members: Volume 13: The Error of Russell, then we will, in Volume 14: The Error of Heidegger, examine the issue regarding both the existence of nothingness and the functionality of nothingness.

Through the examination of Einstein and the inadvertent metaphysical understanding Einstein initiated regarding the significance of 'i', it is now possible to remove a portion of the state of confusion initiated at the beginning of this volume for now we better understand:

Einstein moves our perceptual understanding regarding the Kant/Hegel system being filled with 'timelessness and spacelessness' back into the system being filled with time and space.

As such, ‘time and space’, with the help of Einstein, once again have a location within which they can be found. In addition, the understanding regarding the role of ‘time and space’ and the role of ‘timelessness and spacelessness’, as well as the understanding regarding the interrelationship between ‘time and space’ and ‘timelessness and spacelessness’ no longer remain in a state of confusion. Even more importantly, the existence of such an interrelationship is now recognized as a significant aspect of the ‘larger’ system.

Symbiotic Panentheism[2]

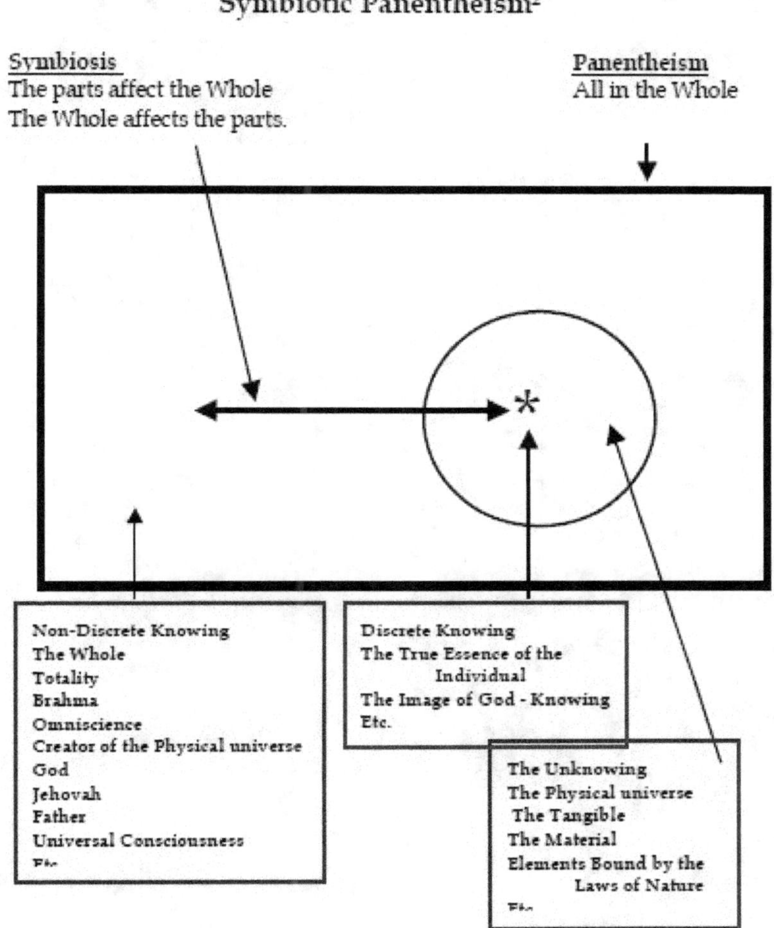

2000 CE

Understanding Reality Evolving

[2] Why add the adjective 'Symbitotic' to the noun Panentheism? There are, as with all nouns, many subgroups of the noun. In the case of panentheism there are many types of panentheism. Within the works produced by the author it is symbiotic panentheism which provides the answers to the the third question: Why? Why does the physical universe exist? Why do we, you and I, exist? Why did Universal Consciousness create Discrete Consciousness? Why was nothingness created? Why have we been unable to resolve age old philosophical, religious and scientific paradoxes and puzzles? Etc.

Preview
Volume 13
The Error of Russell
Resolving the Problem of Non-Members

Preview
Volume 9
The Error of Russell
Resolving the Problem of Non-Members

The Universe

Is a System Filled with:
The Abstract - Zeno
The Physical - AristotleFree Will Immersed Within Determinism - Boethius
Humanity Loses Its Concept of Being the Center - Copernicus
Imperfection - Leibniz
The Void of Time and Space - Kant
The void of First Cause - Hegel
Time and Space - Einstein/Science

Separation Through Exclusion - Russell/Religion

Perfection exists: - Leibniz

The Omni-s - Leibniz

5. Omnipresence
6. Omnipotence
7. Omniscience
8. Omnibenevolence

?????????? - Hegel

Understanding Evolving[3]

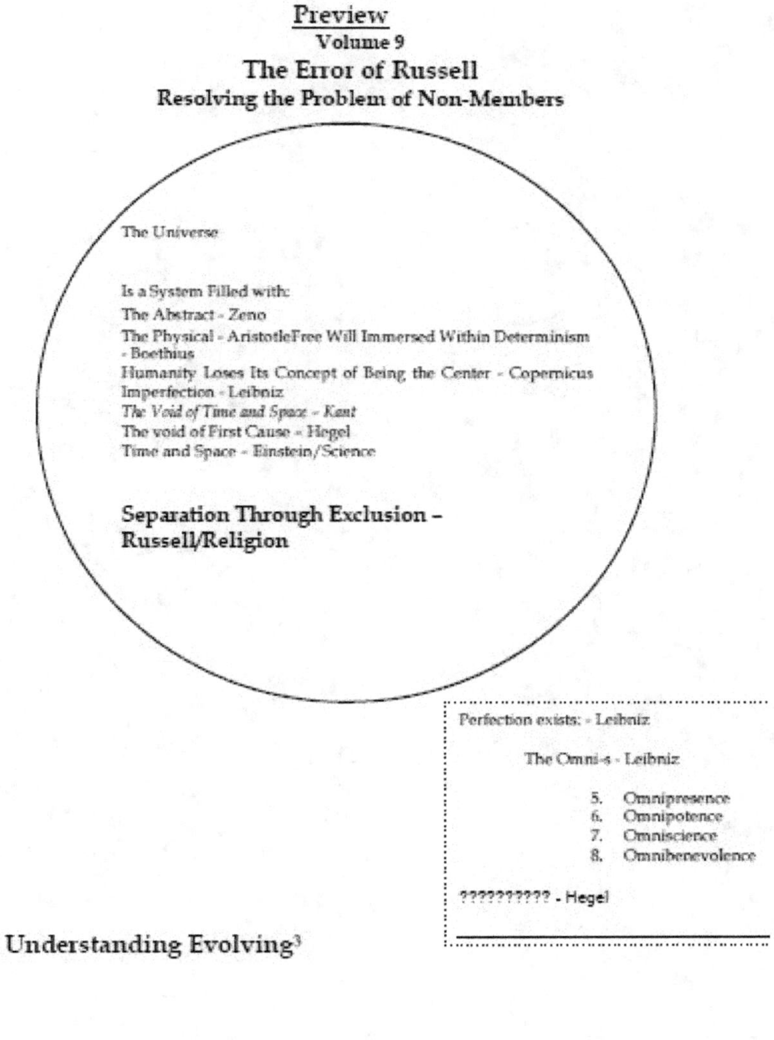

Volume 13

Panentheism

Addressing
The Mathematics
Of
non-Members

Resolving the Problem
Of
Separation Void Exclusion
The Individual Acting Within God
Nothingness Existing With Functionality
Via
Panentheism

Daniel J. Shepard

Channel

Table of Contents

Part I: The Paradox of seamlessness and multiplicity

1. Introduction: 'Nothingness' is an integral part of it all
2. The Wittgensteining of Russell:
3. Understanding Russell's paradox:

Part II: Resolving the issue with a new metaphysical perception

4. An alternative solution to Russell's Paradox:
5. The significance of Russell's paradox as discussed in a simulated conversation between Russell and Wittgenstein:
6. Simplicity itself: The end of the beginning
7. Caution #1: This section is intended only for the mathematically and scientifically inclined
8. Caution #2: This section is intended only for the religiously inclined.
9. Caution #3: This section is intended only for the philosophically inclined

Terms/concepts

Abstract
Abstract Functionality

Concrete
Concrete Functionality
Incrementalism
Illusion
Multiplicity
Real
Real Illusion
Seamlessness
Silent conspiracy of collusion
Singularity of location
Totality/Whole

Volume 13
Panentheism Addressing the Mathematics of non-Members

Separation Void Exclusion
The Individual Acting Within God
Nothingness Existing With Functionality

Part I: The Paradox: Understanding nothing leads us to understanding everything

1. Introduction: 'Nothingness' is an integral part of it all

In the end, the individual acting within God, is simple to understand. Go to the end first if you wish for this is not a mystery story.

This is an essay dealing with the understanding of life and to go to the end in order to understand the beginning is no more a paradox than the famous paradox Russell put forth to philosophers and mathematicians.

This essay will begin where we, humankind, have lead ourselves as we attempted to slash our way through the jungle of life's seemingly endless paradoxes.

These paradoxes, which life has persistently thrown across our path, are signposts for us.

They are indicators that we do not have all the answers.

They warn us to beware.

They warn us there is something wrong with our perception of life; there is something wrong with our thinking.

They have a function of their own, they direct us towards a state of understanding where we are, what we are, and why it is we exist.

In this essay, we will be attempting to understand the likes of Russell, Wittgenstein, Frege, Plotinus, and back again to Russell as we attempt to move past parts of Heidegger.

Our objective will be to step back in time in order to get back to today. What is the point of going to all this trouble just to get back to where we started?

The point is to bring back with us a new perception regarding a simpler solution to Russell's paradox.

Why is this important? Presently we have a solution to Russell's paradox, which involves a complex understanding of 'separation through exclusionism', which in turn represents what we do to people in society.

We separate individuals and groups from our own groups and ourselves. Once having separated them from ourselves we exclude them from ourselves through a process of rejection, exclusionism, and separation.

It is Russell's paradox, which provides the key to rectifying these constant actions of rejection.

For this reason we will accompany Russell as he travels eighteen hundred years back in time. This trip will allow us to bring back with us a different solution to Russell's paradox.

This trip will allow us to bring back a process known as 'separation through inclusion'.

Now the name would seem to imply our creating a paradox to act as a solution to Russell's paradox but as we shall see it does nothing of the kind.

What it does is allow us to find a much simpler solution to Russell's paradox.

'But what does this concept of 'separation through exclusion' as opposed to 'separation through inclusion' have to do with me?' you may ask.

The process provides an alternative means to resolving a fundamental paradox of mathematics, which in turn can be applied directly to the process of understanding life.

It is the simplistic resolution of complex paradoxes, which provides us with a simplistic understanding of life.

It is through this process that we shall see 'Ockham's razor[i] not only cuts away the complexity of science but becomes the primary tool for Husserl's bracketing[ii]. Ockham's Razor now becomes not only a principle axiom for science but now moves on to become a principle axiom of philosophy.

This is an essay beginning in complexity and ending in simplicity.

Why is it that we must begin in complexity rather than begin at the logical point of origin, the point of simplicity?

We begin in complexity for it is through complexity that we presently have begun to understand Russell's paradox[iii].

Presently we have solved Russell's paradox in a complex fashion. This has led us to understanding life in a complex manner. We cannot understand the simplicity of life as long as the basics remain complex.

The solution to this problem lies in the understanding of Russell's paradox.

If you are thoroughly confused, may I suggest you go to the ending of this essay, this mystery, to find the solution.

Then, when you understand, the ending you may find it easier to follow this essay regarding the journey through life we have made as a species.

Once you understand the end of this essay, you will begin to understand why it is that we must go back and make a correctional adjustment to our journey as individuals and as species.

The question becomes, 'How far back in time must we travel to do all this?'

We must go back 1700 years. Who will lead us on this backtracking expedition?

The honors will go to Bertrand Russell himself. Russell verbalized the paradox in 1901.

As such, it is Russell's paradox. Therefore, it will be Russell who will lead our backtracking expedition, which will lead us to an understanding of life.

'What will an understanding of life do for us?'

It will lead us to understanding our purpose for existing while at the same time confront us with an even greater question to which we will sorely want an answer. But more regarding what this question is later. '

And what good will all this 'understanding' do us?' you may ask. Why it will move us one step further along the path of perceptual understanding we have always traversed as individuals and as a species.

So where do we begin? Oddly enough we do not begin at the beginning, rather we begin at the end.

We begin where we have taken ourselves up to today.

We begin with what we have in place, the complex as opposed to the simple, for it is the complex we have established and the simple we have overlooked.

So what was Russell's Paradox?

> *Russell determined there is a set (groups of items) made up of sets whose members (elements) ARE members of themselves. He then concluded there must be a set made up of sets whose members ARE NOT members of themselves.*
>
> *Russell could not explain how a set could exist that is a set of sets whose members ARE NOT members of themselves while at the same time have a set of sets that are members of themselves. The paradox becomes: If these two sets exist, then is the set composed of sets whose elements are not members of themselves a member of itself or not? If it is not a member of itself, then by definition, it must be a member of itself. On the other hand, if it is a member of itself, then it cannot be a member of itself.*
>
> *This paradox is said to provide devastating evidence that logic cannot fully be squared with math. As such, this paradox became a major obstacle for philosophy.* '[iv]

The only way math appeared to be able to resolve this paradox was through a means of 'separation through exclusivism'.

This solution in turn reinforces the concept of 'separation' and 'exclusivism' in our society.

This has lead to the concept of separation, isolation, and rejection within our society, which has lead to an emotional sense of hostility, antagonism, and anger seething beneath the surface of these very groups and individuals being separated, excluded, and rejected within our society today.

This in turn has initiated violent and angry anti-social acts to erupt throughout the social fabric our society (i.e. Littleton, Ireland, Sudan, Afghanistan, and Kosovo).

Separation through exclusion establishes a complex solution to Russell's complex paradox.

There is, however, another means of resolving Russell's paradox.

This second solution, separation through inclusion, lies in simplicity not in complexity and anyone aware of the concept of Ockham's Razor knows exactly what means for the complex solution.

With a new solution to Russell's paradox, comes a new understanding regarding the resolution of exclusionism, separation, and rejection as a concept in and of itself. In essence,

Russell's paradox is a model of life as we presently view it, as we presently live it. As such, a new solution, a new understanding to Russell's paradox leads us to a new understanding regarding life in general, the purpose of life in particular.

With a simplistic understanding of the purpose of life comes a simplistic solution to most of our seemingly complex social paradoxes.

Russell's paradox is a great analogy for understanding why life appears so complex when in fact it is so simple. It is the understanding of a more simplistic solution to Russell paradox, which leads us directly to a different but simpler understanding of life.

It is the understanding of Russell's paradox, which leads us to a completely new perspective of life.

It is for this reason, that solving Russell's paradox has such positive potential for us as individuals and for us as a species. The paradox will be examined in five sections:

1. The significance of the paradox is addressed in a simulated conversation between Russell and Ludwig Wittgenstein.
2. The understanding of the paradox itself is examined in a simulated conversation between Russell and Frege.
3. Using reason, a simplistic solution to the paradox is explored in a simulated conversation between Russell, Frege, and Plotinus.
4. The solution to the paradox is examined more extensively in a simulated conversation between Russell and Himself.
5. Russell revisits Wittgenstein for one last simulated conversation regarding the significance of this alternative solution to his own paradox.

The body of the article is then summarized in four summations.

The four summations are given beginning with the most complex and ending up with the most simplistic.

Russell's paradox is a complex paradox, which can be understood only if one is willing to take the time to do so.

Once one understands the paradox, one can understand the solution.

Once one understands the solution, one can understand the significance.

Why does this preface precede the heart of the essay?

This preface, preceding the understanding of Russell's paradox, was included because as philosophers we must understand, it is not the paradox that is significant but rather it is the lesson the paradox offers us that is important.

The paradox is much like the relationship, which exists between the canary and the miner.

The canary alerted miners of dangerous odorless fumes in the mineshafts. The miners knew that to ignore the warnings of the canary could lead to major injuries if not death.

As such, when the canary appeared weak and falling into a state of unconsciousness, the miners would stop what they are doing, evacuate the mine, and begin a process of preventative maintenance.

Russell's paradox does exactly the same thing for us as individuals and as species.

The philosophers need to repair the danger to which the very existence of the paradox alerts us.

This essay not only suggests but offers a more simplistic alternative to our present day resolution to Russell's paradox.

More importantly, this chapter explores the lesson this more simplistic solution teaches us regarding life.

It is important that we as philosophers and individuals learn this lesson in order that we as individuals and as a species not fall into the same behavior patterns we have generated in the past.

The most significant lesson Russell's paradox appears to offer us appears to lie in understanding that our perceptions lead to actions, which lead to reactions, which lead to an ambiance which descends upon all of us.

Perceptions descend upon us either as a dense, damp, bone chilling fog, or as a warm embracing light.

It is this lesson, brought out in the fifth and last step of this essay, where the significance of Russell's paradox lies.

We, our species, you and I, are on a journey.

We travel in a smaller sense as individuals. In a larger sense, we travel as the species, Homo sapiens.

In an even greater sense, we travel as one of but many forms of awareness within the universe.

Beyond that, who knows, but we are beginning to speculate as to that concept also. Our journey is one taken in time.

It leads from point A to point B. Philosophers are the point people in our journey. Philosophers develop perceptions regarding our journey.

Our journey began a long, long time ago. It is a journey, which has many twists and turns.

Our pilots, philosophers, not hearing anything to the contrary, took many wrong turns.

It was not entirely the fault of the philosophers for making these wrong turns.

The pilots had little information to direct them other than gut instinct derived from the information on hand and the information on hand was primitive relative to today's understanding of the universe.

There is no doubt, however, that it was the pilots, philosophers, with the aid of their two navigators, religion and science, who need to identify at what juncture we made the wrong turn.

It is religious leaders, scientists, and philosophers together who need to adjust our course and get us headed in the correct direction.

Russell's paradox provides us with the clue.

On our present course, Russell's dilemma erupts into a paradox. On a different course, it does not evolve into a paradox.

As such, it is Russell's paradox which acts as the signpost for our species regarding where it is our perceptions of life have gone astray.

The point is, the course we take is our choice and we can change directions if we can determine where it was we took the wrong fork in the road.

The juncture at which we took the wrong turn was during the age of Plotinus (AD 204 - 70).

Perceptions reaffirmed during the age of Plotinus generated actions, which in turn generated reactions.

It was these reactions which generated social ambiance.

It was this ambience, which washed over every one of us throughout the centuries, which were to follow Plotinus.

It was a perception reinforced during the age of Plotinus, which was to wash over us and penetrate every aspect of our society for thousands of years to come.

It was a perception culminating during the time of Plotinus that was to climax in a crescendo of moans emitted from the lips of millions upon millions, hundreds of millions of souls who were to travel from point A, AD 1900, to point B, AD 2000.

This perception was to take on the appearance of being unstoppable until Russell came along with his paradox.

But we did not listen to what the paradox had to tell us. Russell did not listen to what the paradox had to tell him.

We, philosophers, turned our heads and humanity paid the price.

It is time to go back to Russell and reexamine his paradox for it has much to teach us and we cannot afford to ignore the lesson any longer.

Shortly before one hundred million people were to die violently during the 1900's at the hand of man, Gottlob Frege (AD 1879 – Concept-notation) emerged as a lone voice in the field of philosophy.

Frege showed us how to resolve the issues, which were to explode upon the scene of the twentieth century.

Frege, through his development of mathematical logic, accelerated the development of analytical philosophy.

Frege demonstrated that mathematics had the potential to reshape our perceptions regarding who we are and thus avoid the inhumane calamities, which were to consume our century.

Wittgenstein (AD 1889 – 1951) came along soon after Frege and established the idea that philosophy was inept due to language. Wittgenstein introduced the idea of putting all the energy of philosophy into finding a perfect language. Wittgenstein suggested philosophers should place their energies into finding this language rather than examining how philosophers could improve society through perceptual changes.

Then, in 1901 Russell found his famous paradox, which drove home the final nail in the coffin of optimism for our century.

The coffin, filled with the hopes of over one hundred million souls, was lowered into the ground and covered up by none other than Heidegger himself.

There are events in history, which prove to be turning points, pivotal points, junctures where we humans can decide to go one way or another.

Although they may appear to center around one person, they usually center upon a small group of people.

Heidegger, Russell, Wittgenstein, Frege, and Plotinus are just such a case.

These five men proved to hold the fate of millions of men in their hands.

The events, surrounding these five men, represent the key points in history upon which the fate of millions of souls depended. Ironically,

Russell's paradox proved to be the last bastion of hope for the hundred million souls who were to be violently violated in the twentieth century.

As innocent as Russell's paradox may have been perceived when elucidated by Russell in 1901, it proved to be the last hope for hundreds and hundreds of millions of souls.

Souls who were to find themselves emotionally scarred during the most technologically advanced century in the history of humankind.

Russell's paradox was perceived to be a paradox but Russell's paradox is not a paradox.

Russell's paradox is instead the unrecognized Rosetta stone of metaphysics.

Russell's paradox is the key as to how philosophy got off on the wrong path regarding the essence of life and its relationship to the whole.

Russell's paradox holds the key as to why we moved down the path of violence and subjugation for the eighteen hundred years following the observations of Plotinus.

Resolving Russell's paradox can show us how to reverse this trend and move onto the path of tolerance and respect for the individual.

The past is the past.

We must not look at the past and fall into despair. Rather we need to examine the past and learn from it.

We must not study the past in order to learn how to prevent repeating the past.

Rather we must study the past in order to learn how to change the future.

As such, if we are willing to take a fresh look at Russell's paradox we can enter a new age of philosophical rejuvenation.

We can enter an age, which will rescue philosophy from its present path of stagnation and futile attempts at generating new and exciting ideas.

We can propel philosophy into an age of positive potentiality, an age capable of dramatically changing the future.

To do this, however, takes the willingness to admit that something is wrong.

It takes the willingness to admit that we went wrong somehow, somewhere. It takes the willingness to retrace our steps and take an honest look at the journey we have made.

Keep in mind, this process of retracing our steps is vital if we are to reevaluate where it is we went wrong in philosophy.

The error does not have to be a big error.

The spaceship Challenger, a multibillion-dollar space ship, was destroyed in flight because of a 'small' error.

It exploded because of a simple devise known as an 'O' ring.

The 'O' ring would have been ok except for a slight variance of a few degrees of atmospheric temperature.

Had the air temperature been a few degrees warmer, the 'O' ring would not have malfunctioned.

Had the 'O' ring not malfunctioned the Saturn rocket would not have exploded. Had the rocket not exploded, the space ship, Challenger would have completed its mission.

Had the space ship Challenger completed its mission, the seven astronauts would not have died.

Had the seven astronauts not died, the exploration of space would not have been set back several years. This is known as the 'butterfly' effect.

If you do not believe the butterfly effect applies to philosophy, think again, for as we shall see, the butterfly effect pertains even more dramatically to philosophy than it does to science.

In addition, science has learned from experience, that some of the greatest experiments are the ones that are failures for they show us what not to do.

This same concept applies equally as well to philosophy as it does to science. Some of the greatest experiments of science have shown us what perceptions were in error.

These 'failed' experiments demonstrated what perceptions we were to relegate to the annals of history in order to move on and explore a more realistic perception of reality.

The Michelson-Morly experiment was one such experiment, and with it, 'ether' became a concept not forgotten but rather simply a part of history.

Science has much to teach us. However, we, philosophers, are a rather vain creature.

We are a little slow on the uptake when it comes to admitting that the other fields of perception, observation and belief, have significant lessons to offer us, we the supposed experts in the field of reason.

With this having been said, we are now ready to back track through time and look at an alternative trail of events Russell could have taken as opposed to the trail of events he did take.

---

[i] Ockham's razor: William of Ockham (1285 – 1349): Ockham argued that when we are trying to explain the nature of something, we should use the fewest ideas of things as possible. 'The complete Guide to Philosophy', Jay Stevenson, Ph.D., Alpha Books, 1998, p.85

[ii] Bracketing: To set things straight, Husserl says we should ''bracket' all the assumption we have about the world when we experience things, so we'll be able to see past all the layers of meaning that have built up around them. In other words, we need to set these assumptions to one side in order to try out other assumptions. In doing this, we will be able to appreciate new possibilities. 'The complete Guide to Philosophy', Jay Stevenson, Ph.D., Alpha Books, 1998, p.201.

[iii] Russell's paradox will be explained as the essay progresses

[iv] 'The complete Guide to Philosophy', Jay Stevenson, Ph.D., Alpha Books, 1998, p.193.

www.ingramcontent.com/pod-product-compliance
Lightning Source LLC
Chambersburg PA
CBHW072301200526
45168CB00014B/97

* 9 7 8 1 5 0 7 7 4 1 9 8 6 *